Change Your ENERGY Change Your LIFE

Published in 2023 by Welford Publishing

Copyright © Dr. Cornelia Kawann 2023

ISBN: Paperback 978-1-7390970-2-8

Dr. Cornelia Kawann has asserted her right to be identified as the author of this Work in accordance with the Copyright, Designs and Patents Act 1988.

All rights reserved. No part of this publication may be reproduced, stored in a retrieval system, or transmitted in any form or by any means, electronic, mechanical, photocopying, recording or otherwise, without the prior permission of the copyright owner.

A catalogue for this book is available from the British Library.

Editor Christine McPherson

Change Your ENERGY
Change Your LIFE

Dr. Cornelia Kawann

Disclaimer

This book is designed to provide helpful information on the subjects discussed. It is general reference information which should not be used to diagnose any medical problem and is not intended as a substitute for consulting with a medical or professional practitioner.

Some names and identifying details have been changed to protect the privacy of the individuals.

Endorsements

Change Your Energy - Change Your Life is a must-read if you feel overworked or exhausted! Exciting and empowering, the scientific insights offered by the author translate into practical exercises. The causes of exhaustion become evident, and the solutions become both accessible and actionable.

With humor and empathy, Cornelia imparts hands-on advice to let go of old habits that hold us back, enabling the reader to step into their own power. This book holds the keys to the full-energy future we all wish for ourselves!

— Dr. Mara Catherine Harvey, CEO VP Bank Switzerland, Author & Founder SMARTWAYTOSTART.com

I love this book! Cornelia Kawann has created what is really more like a fun, inspiring workshop filled with useful stories, fascinating information, and practical revealing exercises to make the message come alive in your life.

While many books advise writing down notes or thoughts, Cornelia gets us moving and reflecting and experiencing the key principles of this marvelous energy book. Cornelia is no stranger to the field of energy, with advanced degrees and a career spent in electrical engineering. It was only after her body "failed her" that she began exploring just how miraculous is our human energy system. You will be amazed by the miracles you too will discover!

— Bruce Cryer, Author of *From Chaos to Coherence: The Power to Change Performance*, Former CEO HeartMath

Dr. Cornelia opens the door for you to harness the energy within yourself to create change on the outside. Stemming from her abundant well of expertise as an electrical engineer, Dr. Cornelia equips you with energy exercises and scientific insights to go in your personal development toolbox.

If limiting beliefs have caged your potential, let this book be your key. Dr. Cornelia guides you in liberating yourself from the chains of people-pleasing and perfectionism, propelling you toward an empowered, authentic life.

— Portia Booker, Executive Producer

This is such an exciting book! I love the exercises and energy symbols. They will help you get the most 'Juice' out of your life!

Cornelia delivers a clear framework to understand more of who you really are from a scientific and spiritual perspective, and how to use that to have more energy, get healthier, and be more productive. Learning to manage your energy is one of the most powerful things you can do.

— Derek Loudermilk, Author, Scientist, Entrepreneur

Cornelia's book takes readers on a remarkable journey, sharing profound insights between the connection of our physical, emotional, and spiritual well-being through the lens of energy. An essential read for anyone seeking to truly unlock the boundless possibilities of what's within!

**— Farah Nanji, Executive Producer,
Podcast Host, Mission Makers**

As a Personal PR Strategist and Global Super Connector, I've had the privilege of working with individuals who possess that unique spark, that magnetic energy that sets them apart in a crowded world. One of these people is definitely Cornelia.

Her book is a revelation for anyone looking to harness the power of their personal energy to amplify their presence and influence.

In her masterful guide, Cornelia skillfully blends years of expertise with practical wisdom, offering invaluable insights on unlocking your innate energy and channeling it effectively.

As someone who understands the importance of personal branding and global connections, I wholeheartedly endorse this book. It's a must-read for those seeking to not only stand out but also make a lasting impact in today's interconnected world.

Cornelia's approach aligns perfectly with my beliefs in personal PR and global networking, making this book an essential resource for anyone looking to thrive in the modern landscape.

— Jessica Fabrizi, aka JFab, Personal PR Strategist and Global Super Connector

If you are dreaming of living life on your own terms, this book will help you to achieve that. It is a very powerful step-by-step description of leaving your past behind and thriving with more energy. By unleashing your unique energy, you transform into a person that stands out from the crowd. Imagine waking up each day with more enthusiasm, feeling empowered to tackle your tasks, and having the stamina to pursue your goals.

Cornelia not only relates her own process, but also gives you actionable examples of what to do and what not to do. Her life is a testament to the idea that changing your energy can lead to embrace new possibilities you were not previously aware of and take inspired action toward the life you desire.

— Jesse Panama, Founder of UltimateVida.com, Author of *The Art of Freedom: Kiss Your Job Goodbye, Build an Online Tribe, Leave a Legacy*

With this book for businesswomen Dr. Cornelia hits the nerve of the times. It's so important to look after ourselves when we are juggling multiple roles as a businesswoman, care giver, mentor, and so much more.

It is such a waste of energy pretending to be someone else and trying to fulfil expectations. That's why it is so important to manage our personal energy and to bring our authentic self to the table. This makes the difference in our lives and opens up new opportunities.

We can't pour from an empty cup, so it's important we protect our energy by knowing how to manage our valuable energy and having a self-care routine. It's a vital, ever-evolving part of creating work-life balance and boundaries, and this book guides you through this process.

Being a High Energy Woman enables you to let go of the imposter syndrome and know it is ok to let go of fear of failure as it's part of the journey. Expanding your energy will not only make your relationships (be it in business and in your private life) deep and meaningful, but also your whole life.

— Mandy Sanghera, Award-winning Philanthropist, International Human Rights Activist, and Global Campaigner

*To Emma and Ronja
who are always believing in me and my dreams.*

*And to all Energy Beings
who start managing
their Personal Energy.*

Never forget that you and your frequency are unique and special!

First Foreword

In the ever-evolving dance of life, where light seamlessly intertwines with shadow, and vibrancy coexists with the void, there lies an intricate tapestry of tangible realities and unseen forces. Such is the world within us and around us — paradoxical, enigmatic, yet profoundly familiar. "Change Your Life – Change Your Energy," penned by the esteemed Dr. Cornelia, is a profound exploration that invites readers to delve deep into the realms of personal energy — a force that shapes, defines, and propels every facet of our existence.

Dr. Cornelia begins her narrative with a raw and visceral account of the feeling of being trapped in a "dark hole." This experience, though intensely personal, evokes a universal sentiment of uncertainty, isolation, and the innate human quest for self-discovery. How many times have we found ourselves standing alone amidst a sea of faces, grappling with choices and paths that once seemed so alien? For Dr. Cornelia, such moments, shrouded in obscurity, are the genesis of our most profound personal revelations.

The journey of self-realization, as illuminated by Dr. Cornelia, is punctuated with the recognition that many facets of our identity — our beliefs, thoughts, and actions — are not intrinsically ours. They are borrowed, shaped by society, family, and past experiences. Yet, buried beneath these layers of conditioning is our unique vibrational frequency — a signature of our energy that holds the sacred key to our authentic self.

Even though I'm not an expert in personal energy and tend to be more pragmatic, this book, along with the energy exercises inside, guided me in exploring the realm of personal energy safely. Using the compelling analogy of radio waves, Dr. Cornelia masterfully elucidates how our inherent frequency, our truest essence, is continually modulated by the world around us. Within this intricate modulation, she poses a pivotal question: Do we lose our true selves, or do we evolve into a complex mosaic of experiences, memories, and life lessons?

Challenging the traditional paradigms of Newtonian physics that have long been ingrained in our collective psyche, Dr. Cornelia introduces her readers to the groundbreaking realm of quantum physics. In this revolutionary world, she posits that we, and everything around us, are embodiments of pure energy — an energy that goes beyond the confines of what we perceive as solid matter.

Drawing from her vast expertise and her personal spiritual journey, Dr. Cornelia seamlessly bridges the chasm between the tangible scientific and the intangible ethereal. Her writings offer a comprehensive understanding of life, energy, and the vast cosmos. From elucidating the profound emptiness within atoms to acknowledging our expansive personal energy fields, she orchestrates a transformative journey, urging readers to shed the anchors weighing down their souls and to harmoniously attune with their genuine frequency.

But "Change Your Energy – Change Your Life" is not merely a book — it's a clarion call, beckoning readers to embark on the most pivotal voyage one can undertake: the odyssey back to their true essence. As you immerse in these pages, be prepared to introspect, to question, and ultimately, to resonate with the vast symphony of existence.

In Dr. Cornelia's words and wisdom, here's to the profound journey of discovering, understanding, and harnessing our unique energy. Through this exploration, may each reader find their irreplaceable niche in the boundless expanse of the universe. Welcome to the world of energy as envisioned by Dr. Cornelia."

— Dame Tessy Antony de Nassau,
Entrepreneur and Business woman

Second Foreword

Energy Boost For Working Women

How to go from depleted and exhausted to being fully vivacious and having energy like a teenager again

Your personal energy influences every aspect of your life – from how you feel to what you think, how you behave, and your overall vitality. Mastering your own energy is one of the most important skills you will ever learn.

This book is an invaluable guide for working women who aim to conquer life's challenges without succumbing to exhaustion and burnout. It's a treasure trove of practical energy healing tools that can be applied in any life situation, designed to infuse your days with joy, enthusiasm, and the vibrant energy needed to cherish moments with loved ones and navigate life's obstacles with unwavering confidence and skill. You will learn how to influence the energy of your own body so that you can wake up excited with an infectious smile, eagerly anticipating the day ahead.

As you explore the pages ahead, be prepared to be taken on an engaging and enlightening journey through the realms of energy healing and science. A scientist by profession and a healer at heart, Dr.

Cornelia Kawann stands as living proof that science and ancient healing methods can harmoniously coexist. Her journey melds personal experiences with scientific insights and ancient wisdom, inspiring those who often seek the clues and explanations offered by science before embracing and trying it.

Dr. Cornelia Kawann provides a scientific perspective on the critical importance of mastering 'energetic hygiene', shedding light on why so many of us often feel drained and exhausted. You'll gain insights into how to embark on a journey towards renewed energy and vibrancy. Discover life-changing tools that empower you to shield yourself from 'energy vampires', effortlessly regain boundless vitality, and have the energy like a teenager once again.

This book isn't just for beginners; it's also for seasoned practitioners eager to expand their horizons and acquire new tools. By seamlessly intertwining scientific facts and research and practical energy healing tools, Dr. Cornelia Kawann ensures that this book becomes a captivating companion you'll be reluctant to set aside.

As you embark on this journey through its pages, approach it with an open mind, embrace playfulness, and dare to experiment. Remember, you need not master every exercise immediately. Allow yourself the freedom to explore, play, and discover your unique path to its application.

**— Nina Maglic, MSc, Business Coach,
Marketer & Energy Healer**

*"You can trust only one truth in your life.
And this is the one of your ENERGY."*

Dr. Cornelia Kawann

Contents

Chapter Zero
Where do you begin? 1

Chapter One
Falling into a dark hole 7

Chapter Two
You are not who you think you are 17

Chapter Three
**What your energy has to do with
your bank account.** 25

Chapter Four
Make peace with your body 41

Chapter Five
Your body wants to talk to you 55

Chapter Six
**You clean your physical body every day –
How about your energy body?** 73

Chapter Seven
**Immediately recharge your energy
without a magic wand.** 91

Chapter Eight
Make friends with your emotions 109

Chapter Nine
Clean up your thought garden 135

Chapter Ten
Before the sun sets, forgive 155

Chapter Eleven
Let go 169

Chapter Twelve
Connect with your heart energy. 195

Chapter Thirteen
Don't give the energy vampires a chance. 213

Chapter Fourteen
Your energy is the perfume to everything you do .. 231

Chapter Fifteen
Your energy shows you your way 257

Chapter Sixteen
This is the end? No, it is the beginning 271

Chapter Seventeen
Your list of ENERGY EXERCISES 291

Chapter Zero

Where do you begin?

Dear Magical Energy Seeker,

I am delighted you are holding this book in your hands and that you are curious enough to read this introduction. I don't believe that coincidences exist – everything happens for a reason. Therefore, I strongly believe you found this book for a reason. Or better – this book found you. Because you are ready to implement its secrets and change your life.

You have probably already tried many different things to move on in your life; to change things in your life; in fact, to actually change your life. But more than likely, everything proved to be so difficult and took sooooooooooooooo long. I get that – I was there as well. And I got impatient, too. I wanted to see a difference right away and didn't want to wait another five years for my life to improve.

So, now you are ready!

You are ready to try something new!

I can feel it. You are ready to change!

I'm sure you have already read a couple of mindset or life-improving books, haven't you? But let's be honest, how many of the exercises did you actually do and finally implement? Two out of ten? I understand that. In most of the books I read about these topics, I looked over the exercises and promised myself that I would do them later. When I had time. When I was more relaxed. When I finished reading the entire book. During my next vacation. There always seemed to be a good reason why I couldn't do the exercises right away.

And I will share a secret with you – please don't tell anybody – but I never ever came back to fill out the coaching exercises. The empty lines in these books are still waiting for my thoughts and responses. And that's why I stopped reading these books. In the end, they made me feel bad because I never completed the tasks. I never made it. I was never good enough. I felt so stuck and didn't know how to get out of it.

That is the first and probably the most important thing that is different about this book. In here you will find no exercises where you have to write things down. Are you curious now?

There are practical exercises which do not involve you writing anything down. But you have to actually DO them! I understand that this can make some people feel a little uncomfortable, as it is so much easier to just read about things but not actually do them. However, this book is all about doing. It is all about feeling the impact in the given moment. And the Energy Exercises in this book are all energetically (and sometimes even physically) ready to be carried out right away.

So, why should you read the book? What will you get out of it?

Energy never ceases. It only changes form; just as water becomes ice or simply evaporates. That's why it's so incredibly important to learn to be aware of your energy, to feel it, to manage it. To raise it up and expand it. To be mindful of it. It is your greatest treasure. And once you have connected and made friends with your energy, there are amazing benefits you can discover, including how to:

01 Communicate with your body.

02 To make conscious choices.

03 Be aware of your energy.

04 Clean your energy.

05 Recharge your energy.

06 Protect your energy.

07 Let go limiting beliefs and stored emotions.

08 Forgive yourself and others.

09 Re-connect to your heart.

10 Manage your energy.

On top of all that, at the beginning of each chapter you will find a little Energy Symbol. These symbols are energetic reminders. You can draw these Energy Symbols on your hand or arm, and every time you look at one it will remind you to do a certain Energy Exercise. Isn't that cool?

And that's not all. When you have the feeling that you need a certain Energy Symbol drawn on a certain part of your body because it needs more energy – please do so!

Challenge accepted? Are you all in?

Can I count on you?

The most exciting advantage of this book is that you can do these Energy Exercises right now, and I would certainly encourage you to try them straight away. You will immediately feel a transformation, and your energy will literally shift before you reach the end of the book.

Ok, maybe if you are reading this book on the train, or in a coffee shop, you might want to wait until you get home. ☺ But if you are anything like me, and you don't care what other people think, just do the exercises when you need them. Wherever you are. You can even share the Energy Exercise with the person sitting at the table next to you. Just make it a bit of fun and use it to start a conversation. You will be amazed at how things evolve and what happens afterwards.

I am already so far down the road that I do my Energy Exercises everywhere, and that's the beauty of them. You might feel a little bit apprehensive the first time, but by the time you reach Chapter Ten you are going to be doing the Energy Exercises anyway ☺. And honestly – most people don't even notice what you're doing, because they are so preoccupied with themselves.

If you want to read the book all the way through and then go back and do the Energy Exercises, that's up to you. But I would emphasize that the sooner you do these Energy Exercises the better. And the sooner you will be able to change your life! Is it a deal?

Then try them out for yourself! And please share how you get on, as I'd love to know how you feel. And there is a special gift for anyone that sends me a picture – or shares it in my Facebook Group, *Change Your Energy Change Your Life* – of themselves doing any of the Energy Exercises in a public place.

Finally, there is one more thing I feel I have to say. If you are not interested in trying the Energy Exercises, if you're not interested in implementing these powerful life-changing tools, then maybe this isn't the right book for you. Because you won't achieve the changes just by reading the exercises. This is an activity book, so actually doing the Energy Exercises is essential!

Don't think you can just read about energy to expand your energy. Unfortunately, that's not how it works. This book is all about reading and implementing at the same time. And this is something I realized when I was writing this book. I actually needed to do some energy work on myself before I could find the words to be able to help anyone else. So, as you can see, I still have to re-examine my thoughts and my beliefs and go back and do my Energy Exercises on a regular basis. The exercises are very powerful, but you are not going to read this book and be completely transformed and fixed for life. You will find yourself in the same situation as me: when things come up in your life, you can easily come back to a specific Energy Exercise to help you cope.

So, you can put down the book now and never do anything. And that's fine. But then you probably won't live a life full of energy, or have your own unique energy blueprint which enables you to fulfill your purpose in life. And if you don't have enough energy, you will be missing out on many opportunities. Because you just don't see them.

Research has shown that people with low energy have only limited perception. And this is because they are running on power saving mode. On the other hand, with an overflow of personal energy, you can experience all these new and exciting emerging possibilities for success, relationships, health, and wellbeing. And you will have the energy to execute them.

I am 100% sure that you will make the right choice. You will find the balance and do whatever feels right for you, however it serves you best.

Because it is you who is holding this book in your hands. Your energy and this book have already connected – the magic has already begun!

USE it.

energy-on!

Your Magical Energy Guide

Cornelia

Chapter One

Falling into a dark hole

HOPE

"We cannot solve our problems with the same thinking we used it when we created them."
Albert Einstein

Falling into a dark hole

Darkness. There was just this incredible darkness and fear. All around me. And I felt so alone. Alone with all that burden. Alone with the decision of what to do. This is not knowing what to do. How to get out of this situation. It was this indescribable overwhelm that dragged all the energy out of me.

There was a certainty that, yes, I could ask other people for advice. There were people around me to support me. But on the other hand, I also had this overwhelming certainty that it was me who needed to find a solution, to decide what to do next. It was scary to feel so out of control and unable to see what the future held because everything seemed so dark. I had not the slightest clue how to get out of this darkness and to get rid of this fear. I felt completely alone.

But I knew it was me who needed to take on the responsibility to make the decision – nobody else!

Because this is my life.

Because it is my body.

Because it is about me and nobody else.

Because this is nothing you can pass on to anybody else.

I had to solve the issue, but I didn't know where to turn or what next step to take.

How had I got myself into this dark hole? Honestly, I didn't know at that time. I thought I was leading a "normal" life, with everything going along quite nicely. Life was going quite well, even though I have to admit not everything was perfect.

Have you ever felt that you were leading a nice life? Not perfect – but was it ok for you? And then something completely unexpected happened – out of the blue – and you fell into a deep, dark hole, and your life that was once calm and in control was now spiraling? It was so deep that you couldn't see any light at the end of the tunnel?

I had a very good education, a good-looking and nice husband (of course not always, but in general – I mean the nice part), two wonderful little girls, and a demanding management position in the electrical industry where I could make an impact. During my electrical engineering studies, I had learned how to get my way as a woman in a man's world. I always needed to be better than my colleagues and to prove myself. That's why I wanted to show the world, and probably also myself, that I could be everything: a perfect manager, a perfect mom, a perfect spouse, and lead a wonderful life.

And from the outside, it looked like everything was perfect. Goal accomplished!

But behind the scenes, things were completely different. Every morning was the same. When the alarm clock went off, I felt exhausted and even more tired than when I had gone to bed the night before. I had no energy at all and just managed to get through my day, ticking off items from my To-do lists, trying to meet all the obligations from my demanding corporate job and at home from my two little girls, the household, my husband, and from the community. I was stressed out, had numerous physical problems, and had no energy at all. But I thought that was just the way life was.

Then suddenly, quite abruptly and with no warning at all, I found myself in this dark hole. And I had to admit to myself that I had failed. That I had failed dramatically.

It happened to me on a very nice sunny Friday in July 2016. During a routine health check-up, they found a severe health issue. And from one moment to another, my whole world just broke apart and I completely panicked.

I couldn't understand why this had happened to me. I had always tried to meet all expectations, following the mantra "be a good girl and you get what you want". I had tried to do everything right, so why had this happened to me? Why was life punishing me this way? What did I do wrong? These were the jumbled thoughts that were racing through my head as I tried to make sense of everything.

I was a strong, capable, confident woman who suddenly didn't recognize her own reflection anymore.

I realized that everything had changed and that my future was going to be a very different one than I had envisioned. I questioned everything. What was my life going to look like? Would there be a future at all?

I had no energy left. I was overwhelmed by my feelings, my fear of the future. Of everything. Not knowing what to do. Not knowing how to find the light out of this tunnel. I didn't know what the future held for me, and at that moment I was concerned about my family.

I started to worry about my daughters.

I wanted to be there for them.

I wanted to see them grow up, to accompany them on their way and to be part of their lives. It broke my heart that I might not share the happy memories of my children growing up.

As a mother, there's never any doubt that you won't be there. You just always expect that you will be there, because that is your role, isn't it? To nurture and to love and to protect. The mother is the main pillar in most families, and if that pillar breaks away, the whole system becomes imbalanced.

My family was shocked when they learned about my health problem, and of course there was a lot of fear and anxiety. So, I didn't only need to deal with my own fears and worries but also with those of my husband and my mom. As a result, I couldn't really talk to them about how I felt, because they were already running nuts with their own fears.

It's only now that I feel I can share my feelings and innermost thoughts here with you. Thoughts and feelings I haven't even told my own family members and friends, because I felt like no-one else understood me. I am one of these people that needs to figure things out first on my own before I can actually speak about them with somebody else.

That momentous Friday, I realized that a lot of things in my life were not the way they should be – or how I wanted them to be. I was not leading MY life. I wasn't living what was important to me. I had forgotten about a very important parameter – the most important parameter in this equation – and that was myself. My body had been continuously sending me signals that I needed to change something, but I hadn't understood or listened to these warnings.

But that day, I woke up! I knew then that I needed to change a few things in my life, even though I didn't know which ones, or how. I was scared and unsure what to do, but I knew for certain that I needed to do something, because my goal was – and still is – to see my girls grow up and to be with them.

And that was the moment when I started my journey of getting my life and my health back on track. To find out what is good for me and what I want.

As an electrical engineer I know that there is always more than one solution to a problem. And to find a solution to get out of my situation, I started digging into different possibilities. After some research and reading various books, I discovered the world of quantum science and energy work, which opened a door into a completely new world for me. A world that proved to be the light in my tunnel!

Quantum physics is always seen as something very abstract and scientific – but no; it is real life. I liked the idea that – in a nutshell – as each of our cells consists of atoms, everything is just energy vibrating at a certain frequency. This approach resonated with my electrical engineering heart and made it jump. It changed my perspective and made me look at everything just as energy. Every thought, every word, whatever you do, your relationships, your health, and yes, also money – all just energy. From that moment when I understood that everything is energy – which is the baseline of quantum science – and that everything has its frequencies, my life changed completely.

I realized that I needed to solve my problem on a different level – on the energy level. And that was when I found an energy healing method based on quantum physics, working with frequencies. I am an electrical engineer, so this was perfect for me, because I could understand how it works. This was my solution!

And that's how I started to manage my own energy in my daily life. I became more and more aware of my energy, learning who and what sustained and drained it. I experimented and developed tools and techniques on how I could increase and expand my energy, how to remove blockages, and how to protect myself against energy vampires. In short, I expanded and transformed my energy.

Nowadays I am healthy and buzzing with energy. Every morning I jump out of bed ready for the day. I have advanced my corporate career to an executive management position, and despite more responsibility at work – especially now with the current energy crises – I have more time with my kids. And that's because in everything I am doing, at work or at home, I am much more efficient and productive. I have so much energy that on top of everything else I even started my own business – something I had always dreamt of. Can you imagine?

For many people – maybe you, too? – time is their most valuable asset. In my case, I figured out that my energy is my most important currency. After all, what can you do with all the time in the world when you feel miserable or sick?

Although I am sharing my story, this is not about what pulled the ground out from under my feet. There might be something in your life that pulls the ground out from under you, and that could be a parent on the verge of a breakdown, a burnout, a divorce, or even an accident. Everything that is in your life has something to do with you, whether you like it or not. But the good thing is: if you take responsibility, you can change it. And I imagine that if you weren't up for changing something in your life, you wouldn't be holding this book in your hands. Right?

What's important is to remember that you always have a choice. I just was not aware of my choices at that time. But now I know that there is always more than one solution.

And that's why I'd like to invite you into my world of energy. The world that helped me to get out of my dark hole. The world that transformed my life. And into a world where you don't feel alone, anxious, and overwhelmed when emotional complications come into your life. And the great thing is, you don't need to go through the same trouble I did, because I am offering you a shortcut by teaching you about your energy and how to manage it.

I have learned so much about energy management and how much easier many things get when you do them with the right energy or by using your energy in a smart way. Your energy is already there, waiting for you. It is the key to your wellbeing, and you just need to know how to use it.

Nowadays, we cannot imagine a life without electrical energy. And it's the same with life energy. We just need both! To put it bluntly – there is no life without energy!

The purpose of this book is, in a way, to help you realize that all your thoughts, your emotions, and your beliefs are energy, and they all influence your energy. But if you know how to manage them, then you will have all the energy to deal with the challenges of life in a more resilient way; all the energy to live more of the life you actually want to live.

Your Energy Insights

Life is about constant learning, and we can never escape its up-and-down waves and curveballs. However, when you learn how to manage your personal energy, you can surf the right waves more consciously and be less easily knocked down. Your energy is the key!

Chapter Two

You are not who you think you are

YOU ARE ENERGY

"If you want to understand the Universe you need to start thinking in energy, vibrations and frequencies"
Nicola Tesla

You are not who you think you are

There comes a certain time in life when we all ask ourselves this question: Who am I really?

Have you already asked yourself this question? If yes, then congratulations! You are already on the right track.

If somebody points out that something you did or said was not ok, you can always shrug your shoulders and reply, "Sorry, but this is just the way I am!" But isn't that a lame excuse? Are you sure this is the way you really are? Do you think you are all the beliefs, thoughts, and feelings you are having all day long?

What I have learned so far on my journey is that I am not so sure anymore that I know the real me. When I look closer, I see that many things I am doing, thinking, and believing, are things I have learned, observed, taken on, picked up, or were told by somebody else. Most of this happened even unconsciously when I was still a child. But all these thoughts, behaviors, and emotions have already been in my life for such a long time that they feel a part of me. They seem to belong to me. But are they my real me?

Let me share my world of energy with you

From my perspective, we are all born with our beautiful, particular, and unique core frequency. You may ask, what does this frequency have to do with me? Quite a lot, actually.

If we are – according to quantum science – all energy, then it also means that each of us is vibrating at a frequency. A frequency is a measure of how often a vibration repeats. The unit for this is Hertz (Hz). For example, our electrical energy at home vibrates at 50 Hz.

Hence, if we are all vibrating at a specific frequency, from my point of view your frequency is similar to your unique fingerprint. So, there is nobody else but you who has that specific frequency. I would even go so far as to say that it's not the human being who has the unique core frequency, but its soul. This means that each person has its very own original frequency, and you can see your unique frequency as the real you.

The graph below is an example of how your frequency may look when you are born.

Just plain and sparkling like a marvelous diamond.

However, during our lifetime we take on thoughts, beliefs, and energies from other people, and this is like adding layers onto our own frequency.

Let me explain with an example. For me, as an electrical engineer, it's a bit like listening to the radio. Can you remember your physics class? Do you know how a radio works?

What listening to the radio has to do with who you are

It's actually quite simple. Sound waves have a limited range. For example, when someone plays the piano or shouts, these are sound waves with a short range. And to be honest, it's often a good thing that not all sounds can be heard over long distances. When you have thin walls in your apartment, it is good that nobody can overhear your conversations.

Nevertheless, in the past people wanted certain information to be audible over a larger range and longer distances, as for example with national radio stations, so that everyone in the country could hear the same music. I know that's hard to imagine in today's times, with the internet and all kinds of streaming services, but that's why big TV and radio stations were built.

There, the information such as music or news is modulated onto a carrier frequency with a large range. If you then tune your radio to this carrier frequency, e.g., 89.2 Hz, then your radio filters out exactly this carrier frequency again, and you hear the actual transmitted information, namely the music or the voice of the newscaster.

Isn't that super cool? And it all works even though sound and radio waves are invisible, and we can't hear certain frequencies.

So, your very unique core frequency you were born with is quasi your very own, pure, and unique carrier frequency. And this frequency changes depending on what you experience and how you live.

This means that in the course of your life many different pieces of information have been modulated onto your very unique carrier frequency. Every thought, every feeling, every belief, every event, everything we carry around with us has a frequency. Some of them have high frequencies, some low. So, nowadays, your actual frequency is no longer plain and sparkling, but can look something like this:

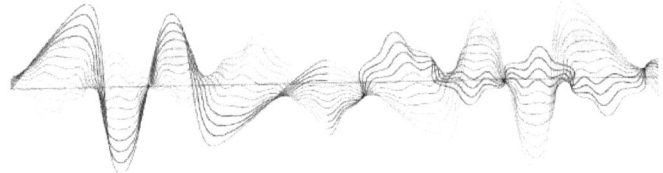

What frequency you are vibrating at depends on your core beliefs, what you think, and how you feel.

Many of these frequencies, however, were modulated on your unique and pure frequency when you were little. This means:

01   **You cannot remember how your unique frequency felt.**

02   **You have been living for such a long time with all these overlaid frequencies as thoughts and beliefs that you think this is you.**

The journey back to yourself

As all this stuff is not really you but has been put on you – or you put it on yourself – the good news is that you can also remove it. The first step is just being aware of it. Afterwards, you can just let it go, get rid of all these wrong frequencies, and go back to the unique frequency you were born with. Does this sound good to you?

I shall be sharing how you can do that with powerful Energy Exercises further on in the book. So stay tuned!

At this stage, the important message I want you to remember is:

01 Be aware that you are energy and that you have your very own and unique (carrier) core frequency.

02 In the course of life, you add through experiences, thoughts, beliefs and feelings a lot of additional information and thus frequencies on your unique core frequency, resulting in a new resulting frequency. This resulting frequency may change continuously.

03 Important here: This resulting frequency is not YOU!

04 From my point of view, the goal is to gradually let go of these modulated frequencies and come back to your original unique core frequency.

05 The closer you vibrate to your very own unique core frequency, the better you feel on all levels, be it physically, emotionally or mentally.

Personal Energy Management is the journey back to yourself and to your true self, which means being fully connected to yourself – not only when you are meditating and doing well, but in your daily life. This doesn't mean, of course, that you are going to have a perfect life, or that you will always be nice and kind. When you are at work, when you are with your kids, when you are with friends, when you are shopping, when you are faced with challenges, the question is: who are you then?

I want to help you be connected to yourself in those moments, being in trust in those moments.

Allowing yourself to be fully you. Being unapologetically yourself. To embrace yourself with all that you are and navigate through life with it. And allowing this not only for yourself but also everyone else around you.

Your Energy Insights

This is what I love so much about Personal Energy Management. When you are working on your mindset, it is all about overwriting a certain behavior and becoming somebody else. With Personal Energy Management, it is all about letting go of everything that is not you and doesn't belong to you, and uncovering your *true self*. Letting go is so much easier than becoming somebody else. And it is not even necessary, because *your true* YOU is right.

Chapter Three

What your energy has to do with your bank account

BALANCE YOUR ENERGY

"Should you find yourself in a chronically leaking boat, energy devoted to changing vessels is likely to be more productive than energy devoted to patching leaks."
Warren Buffett

What your energy has to do with your bank account

"Cheeky Disclaimer"

Just a warning to let you know that I am going to be sharing some scientific information every now and then, but I promise it won't be boring and that it will help your mind to understand the concept of Personal Energy Management so you can transform yourself and your life. Maybe now you are thinking that you can always skip these parts. You can, of course. But then your mind might resist following the Energy Exercises. And the exercises are the secret to unlock the changes you desire.

But if you stay here and remain open-minded, it will help you to get through the process faster. And that is what you want, right?

Why we are all a little bit scared of energy

Energy is so not tangible and hard to understand for most of us, as we grew up in the world of Newtonian physics. It became the cornerstone of science in the seventeenth century and describes a set of physical laws that affect the motion of bodies under the influence of a system of forces. Newtonian physics perceived the Universe as a sort of clockwork model, and even us humans were simply seen as complex machines. Only what could be perceived with the senses and measured by scientific instruments was real. This means, in simple words, if:

01 **You cannot see it**

02 **You cannot touch it**

03 **You cannot measure it**

… then it doesn't exist. Growing up with these laws is the main reason why we are so afraid of things we cannot see. Especially when it comes to energy. Being brought up in a world in which only matter is significant has made us all more or less materialists, because we are so focused on people, places, objects, things, and time in our environment.

In the 1900s, with the beginnings of quantum physics, beliefs changed, and this new science completely revised the way we looked at the structure of atom models. In 1909, Ernest Rutherford's team at the University of Manchester worked on an experiment in which they projected alpha particles (positively charged particles) onto a gold foil. Almost every alpha particle passed it as if there was nothing there at all, making an exception for 1 in 10,000 or so. Hence, Rutherford and his team concluded that atoms are almost empty spaces with a highly concentrated dense mass called the nucleus. Just to give you an idea of the dimensions: If an atom is equal to the size of a football stadium, then its nucleus would be near the size of a football. That's a lot of empty spaces, isn't it?

But actually, it is not empty – this space is in fact energy! That is exactly what quantum science agrees to stipulate, that the Universe, including us, is made up of energy, not matter. In reality, the atoms that form objects and substances that we call solid are actually made up of 99.99999 percent energy and 0.00001 percent matter.[1] And this 0.00001 percent matter also consists of energy, which means that we consist of 100 percent energy. The energy that makes you is the same energy that composes trees, rocks, the chair you are sitting on, and the phone, computer, or tablet you are using to read this book.

They are all made of the same stuff – energy. Now you may ask yourself, if an atom is 99.9999 per-

---
[1] Breaking the Habit of Being Yourself; Dr. Joe Dispenza; 2013.

cent energy and only 0.00001 percent physical substance, why draw all the attention to matter when it is only such a small part of the "physical" world?

Looking at everything from the perspective of energy was the most fascinating thing I've learned in my life. And that was actually after I had finished my studies in electrical engineering, where I got the impression that I already knew everything about energy! It opened a completely new world to me – *the world of energy*!

Electric energy and your personal energy: Is it the same?

As already mentioned before, we all grew up with these Newtonian filters. However, when you are honest with yourself, we have actually already moved away from them a bit. Nowadays, we believe in electric energy, in radio and mobile phones, in W-Lan, although we (mostly) don't see it. Actually, we cannot even imagine a life without these invisible energies, right? So why not allow all the other energies to be part of your life as well?

Yes, I agree:

01 **You don't see energy.**

02 **You don't hear energy.**

03 **BUT – and this is the big but – because we all ARE energy – you can FEEL energy!**

Or let's put it this way: most of us cannot see or hear energy (if you are not some kind of medium).

I bet, though, that you can feel energy – even if you currently aren't consciously aware of that. Yet!

Unfortunately, because we were trained not to focus and trust our feelings but only our mind, we kind of forgot how to do that.

By looking closer into this new energy world, I discovered that there are many similarities between electrical energy and our life energy. For example, from our physics class we remember that electrical energy flows where there is the least resistance. This is similar to our personal energy. When we have an energy blockade, the energy flows in other directions and these areas receive less energy.

Like electrical energy, we cannot see and touch personal energy either, but we can all FEEL it. And we all love the results of the use of electrical energy. We also love the feeling when we are buzzing with energy, don't we?

When I continued working with personal energy, the one thing missing that puzzled me as a scientist was that there seemed to be no unit used to show the positive effects of personal energy work. This was kind of strange to me. We all know that we can measure electrical energy in kWh, so why is there no way to measure personal life energy?

However, after some research, I eventually discovered that there is a unit to measure your personal life energy and the effects of energy work, and it is called Bovis. I shall be sharing more about this in a later chapter, but for now I want to mention how important it is not only to feel the results when it comes to Personal Energy Management but also to measure them.

And on top of that, having the possibility to *measure* your personal energy means – according to Newtonian physics – that personal energy exists.

What your energy has to do with your bank account

According to Traditional Chinese Medicine (TCM), you are born with a certain amount of Qi – your personal energy – with which you can live 100 to 120 years. Sounds great, doesn't it? Don't we all dream of longevity and still want to be healthy when we reach our hundredth birthday?

Assume that your personal energy is a little bit like your bank account – so your personal energy is your starter capital (energy). And just like with your bank account, it is up to you to grow your energy or go into debt. It is up to you how you manage your energy – in the same way as you manage your cash in your bank account.

In the bank, you have some fixed debits and costs which you have to cover every month, e.g., for your rent, food, insurances, and everything you need for your daily life. And you go to work or run your business to obtain your salary, with which you can cover these continuous streams of money flowing out of your bank account.

The same happens with your personal energy. Your day is stressful. You breathe in stale office air; you breathe shallowly because you are stressed. You are in a hurry, and there are still so many things on your To-do list, so you eat reheated convenience food which contains almost no energy and a pizza on the way home. A few coffees got you through the day in good spirits, but you are looking forward to a lazy evening on your couch in front of the TV, so there is no time for exercise. Afterwards, you fall asleep late, but your sleep is restless.

On such days you are not nourishing your energy and there is no inflow in your personal energy account. This is not initially a problem, because your body simply helps itself from its storage, which means in this case that the energy has been depleted from your energy account. This happens automatically without you noticing it. If you are lucky, your body will tell you that its reserves are too depleted or that your energy is deficient. According to TCM, the age we eventually reach depends largely on how well we nourish our personal energy in the truest sense of the word, and how evenly our life energy can flow through our body.

With our busy lifestyles, we often spend or use energy that we don't actually have, continually depleting our reserves. And while you can do that for a short period of time, over a longer spell your body will eventually present you with the bill. Especially if you go far into the minus range, which can usually not be balanced out again so quickly.

Is this perhaps something you recognize from your own experience? Are you working for long hours, staying up late to get everything done, not making the right food choices, burning your candle at both ends? Then when you are finally on your way to your long-deserved vacation, all of a sudden you become sick. And you feel frustrated and question why that should happen to you right now!

When you see me now, you see a person who is healthy, vital, fit, and full of energy. But I can tell you, there were times in my life – not so long ago – when I was totally the opposite. I was tired, I was exhausted, I had continuously different health issues, and then I fell into a deep, dark hole.

Everything I needed to do involved such a big effort. And that was the moment when I realized that in each second, whatever I am thinking or doing, I am trading my personal energy against something else. That's why it is so important that you take care of your personal energy account and are aware of what you are using your energy for. Every single day!

And that's how I started using Personal Energy Management in my daily life. I experimented, healed, increased, protected, and shifted – yes, simply started managing my own energy. And every time I measured the changes. This helped me to keep track of my activities and also of my achievements. On top of that, it held me accountable. If you have lots of energy, you feel better and can manage many things, but you can also prove this by measuring your Bovis values. I will show you how you can measure your Bovis values in the next chapter. Stay tuned!

ENERGY EXERCISE #01 – Become aware what drains and sustains your energy

I know I promised at the beginning that this book is all about doing Energy Exercises and requires no writing. However, this one time for this Energy Exercise, I need to break that rule. Of course, you don't have to write anything, but I can tell you that having an overview of your energy balance in hand can be very useful and supportive in shifting your energy. Check it out for yourself!

Write down all the things – activities, people you know – that you feel are taking energy from you. For example, you know that endlessly scrolling through social media doesn't make you feel good, nor does finishing a bottle of wine when you should have gone to bed an hour earlier, or spending time with a certain person leaves you drained. I understand that perhaps you cannot completely avoid seeing this person because they are a family member or a close friend. But just knowing that this conversation might drain your energy will help you to limit the time you spend with that person. It will help you to stay calmer, and keep your energy protected and with you, because you will be prepared. In Chapter Thirteen you will learn how to do this with a powerful and fun Energy Exercise. Don't miss out on it!

Now, write down all the things – activities, people you know – that you feel give you energy. Like for example going for a walk for 15 minutes, listening to a meditation, taking a warm bath, or treating yourself even to a nice massage. In Chapter Seven I will share an interesting Energy Exercise with which you can recharge your energy immediately and bring yourself to a joyful state. I am so much looking forward to sharing this powerful exercise, and I know it is going to make you expand your energy and open your awareness to so many opportunities.

It doesn't matter how long your two lists are – especially not the positives list! What is much more important is how often you are doing these activities, or they are happening to you per day/per week, etc. And trust me, sometimes it is just nice to have your positives list at hand when you feel down. In those moments you might struggle to think about things that cheer you up, and that's when you can pull out your list and pick one of these activities that give you energy.

I'd like to point out that this whole process is not about making you feel guilty about your life in any way. In fact, it is quite the opposite. It is to raise your awareness and help you to make better choices about what puts energy into your personal energy account and what causes withdrawals.

ENERGY EXERCISE #02 – Check out your energy balance

Now that you know what is giving and taking your energy, we are going to build on it with another Energy Exercise which will make things even clearer to you.

If you really want to see your energy balance, you could even set up an Energy Balance Sheet as shown in the example below, and indicate each activity with a value, specifying how much energy it gives you or costs you, plus the frequency, resulting in a total amount of energy per day. To keep track and have even more awareness about your energy balance, you could fill out your Energy Balance Sheet every evening before you go to bed. You will be amazed how it is going to change once you start keeping track of it.

Energy Balance Sheet

Date	Activity	Value +/-	Freq	Balance
	Drinking coffee	+50	3x	+150
	Cleaning bathroom	-90	1x	-90
	Having lunch outside	+50	1x	+50
	Stressful Meeting	-50	1x	-50
	Meditation	+70	1x	+70
	Picking up Kids	-30	1x	-30
	Swimming 1km	+100	1x	+100

As you can see from the example, there is no right or wrong! Trust yourself when you think about an activity then put in exactly the number that comes up as a first thought. It is the right one. And also, be aware that some activities that give you energy might be perceived as energy draining by somebody else, and vice versa. But with this simple Energy Balance Sheet, you have the perfect (daily) overview of whether you are recharging your energy more than you use it.

As a first step, imagine what a typical day looks like in your life. Think of the many activities and people you spend time with, and rate whether each one gives you energy or takes it from you. It doesn't matter how long your Energy Balance Sheet is. If there are activities on it that recharge your energy, that list cannot be long enough.

This Energy Exercise supports you to connect with your personal energy and learn where it is actually flowing to. These are the three main lessons you learn from your Energy Balance Sheet:

01 **AWARENESS:** Learn to feel your own energy and get to know its level.

02 **FLOW:** With this knowledge you get the daily energy flow a little under control. Similar to your bank account, you can shift your personal energy account from suuuuper draining to something consciously more balanced – in the very best case, even net-positive.

03 **STOCK:** And by the way, you might not be managing your daily flow/bank account at all, with all the energy you actually have available, because your original capital is not even 100% with you anymore. And not because you spent it in your daily flow, but because you "misplaced" it outside of your bank account by "unconsciously lending" it to people, places, and memories.

But you can conjure it all back immediately with the following Energy Exercise.

ENERGY EXERCISE #03 – Get your energy back

Due to our busy lifestyles, for most of the day we are focused on the outer world, on things that are happening around us. We give our energy to people that need our support, we leave it a special place with which we have strong memories, and we put it into things we cannot change at all. All these activities withdraw energy from your personal energy account, but because it is YOUR energy, you can always get it back.

That's why it is very important to call your energy back. Try this very powerful Energy Exercise:

01 **Close your eyes and set the intention to connect with your energy.**

02 **Set the strong intention to call back all the energies that belong to you. Hold this intention to collect all your energy to you.**

03 **Imagine that all your energies, wherever they are, are coming back to you.**

04 **Set the intention to clean your energy before you integrate it again in your energy system.**

05 **Be aware of the difference in your energy and that you feel whole now!**

As we tend to lose our energy over the course of a day, it is good practice to call your energy back every evening so that you feel whole before you go to bed.

So, is working on my physical energy all I need to do?

While experimenting with my own situation, I noticed that even when I was putting a lot of effort into increasing my physical energy, like eating healthily, exercising regularly and sleeping well, I still didn't feel full of energy. There had to be more behind it. It took me a while until I realized that there are different levels of energy!

5 levels of personal energy

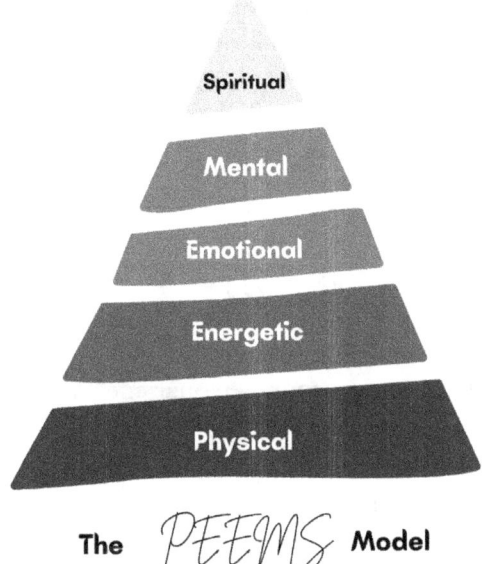

The *PEEMS* Model

As you can see from the graph, your personal energy consists of five layers. There is not only a **p**hysical level of energy, but also an **e**nergetic, **e**motional, **m**ental, and even **s**piritual aspect of your energy. That's why I call it the *PEEMS-Model*.

The most fundamental source of energy is your physical, but the most significant is your spiritual energy!

The first step is to raise your awareness for the different dimensions of your energy. This awareness will trigger interest and action, and you will soon find that managing your energy, not your time, becomes a natural focus for you. Each dimension of energy has a core quality. Understanding these core qualities is essential in order to develop effective strategies for boosting and focusing your energy. And only when each of these levels is full of energy is your personal energy whole. This is when the magic begins to happen.

Let's discover together what these other energy levels are and how they work together.

Your Energy Insights

Being aware of your own energy balance is the cornerstone of managing your personal energy. It is where everything starts. That's why it is so important to be aware of what actually drains or sustains your energy. In the course of time, and when you continuously manage your personal energy, you will notice that the things, moments, and people that drain your energy will change.

What does this mean? Some of them will fade out of your life, and some of them will stay. And new ones will come into your life. But because your personal energy will have changed, you won't *perceive* them as draining anymore. You will give them a different value in your energy balance sheet. And the same may also happen with things that give you energy. The best scenario would be for some things not to have any impact on your energy bank account anymore, as that means you will have become much more resilient and able to keep your energy to yourself.

Chapter Four
Make peace with your body

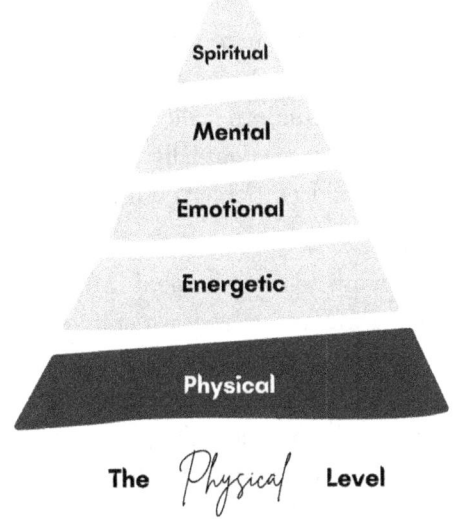

The *Physical* **Level**

As I already pointed out in the last chapter, this book is all about getting to know your different energy levels. The energy you are most aware of is your physical energy. It is the most fundamental one, because it is the foundation for all the other energy levels and the basis of your wellbeing.

Let's get to know your physical energy.

CONNECT WITH YOUR BODY

"Even though the body appears to be material, it is not. In the deeper reality, your body is a field of energy, transformation and intelligence."
Deepak Chopra

Have you ever felt let down by your body?

Anyone who meets me now would never guess that I was a bit chubby when I was little. At least, that was what some other children called me. Looking back now, I don't think it was actually true. However, that was how other children perceived it because I was not as slim as them.

That's why I have looked after my weight throughout my life and why I am still very conscious about my health. I see my body as my vital engine which I need to maintain very carefully. For me, food is something with which I fuel this special engine. So, if I want to keep it in good condition and running smoothly for a long time with only little maintenance work, I need to be very careful about what I eat.

It's why I decided to no longer eat meat when I was 18, I never took any drugs, never ever smoked – not even a single puff – and hardly drank any alcohol. As I got older, I started avoiding white sugar and even sweets in general. During my life, I must have eaten tons of fresh vegetables, salads, and fruits, because I just love them!

Of course, I was aware that a healthy diet is only part of the secret to longevity. Hence, I was always making sure I exercised regularly, especially in summer when I tried to swim a minimum of one km every day. And I always tried to ensure I got enough sleep. As you can see, I really cared for my body to keep it in perfect health.

And then my body let me down. Out of the blue. By just not functioning the way it should anymore. After all I had done for it! I cannot tell you how betrayed I felt. I had sacrificed a lot of time, effort, and energy to look after myself, and then I got this in return? I just couldn't understand what I had done wrong.

However, by looking closer at the situation, I realized:

> YES, I was doing everything that I *thought* was good for my body.
>
> BUT: Do I *enjoy* doing it? And even deeper: *Do I love my body?*

These were tough questions for me at that time. Was I really doing all of this to help my body to function optimally, or was it more to look and feel as healthy as society expected me to?

It took me quite some time to admit that I had never really loved my body. Sure, I had cared for it, but I had never really loved it. I wanted it to function properly so I could do all the things I wanted to do. But I had seen it as an engine that I needed to take care of to get the most out of it. (Wasn't this very Newtonian of me?) However, I had never really accepted and appreciated the way my body looked, and I realized I had always wanted it to look a little bit different. In general, it was ok, but wouldn't it be great if my legs could be longer, and my eyes bigger? And wouldn't it be amazing if my belly were tight and strong, and I would just look the way I wanted to look?

Did you have such thoughts about your body? I bet you did! We all do, right?

Most of the time we focus all our attention and energy on things that are not how we want them to be. I had one of my biggest AHA moments when I saw a masterclass from motivational coach Biyon Kattilathu. In that video he held a white sheet of paper with a black dot in the middle up to the camera and asked: "What do you see?"

Automatically, everybody answered, "A black dot on a white sheet of paper." Then he posed the next question: "What is the percentage of the white part versus the black dot?" We all agreed that it was about 95 percent white versus 5 percent black – or maybe even less. Finally, he asked the last question: "Why do you focus on the black dot when it represents only five percent?"

And that's a valid point there, isn't it? Why do we always focus on the few things that are not perfect and the way we want them to be? It takes so much effort to improve your weaknesses, but only a little energy to thrive on your strengths. If you only polish off the weaknesses, they will all be average in the end. So, it is much more powerful to focus on your strengths, on what is working fine in your life, and on what you love about yourself and your body.

Let's focus on your strengths and beauty spots.

ENERGY EXERCISE #04 – What do you love and appreciate about your body?

Focus on the parts of your body you really love. For example, I really love my feet. They are so well- formed, and I just love looking at them. I am also very grateful for them because they have already taken me to so many beautiful places. They are just amazing! And of course, I love many other beautiful parts of my body, especially my natural curls and my dimples when I smile.

How about you and your body? Take a few minutes to look only at the spots you like, those you are proud of. Your eyes, your neck, your hair, your dimples, the curve of your body, your legs, your décolleté, your breasts - you have so many beautiful parts. Show them your appreciation and your love.

As a next step, think about all the things you are especially grateful for that you experience through your body, for example:

01 **I can experience and feel love.**

02 **I can perceive touch.**

03 **I can feel joy.**

04 **I can climb mountains.**

05 **I can swim in the ocean.**

If you do this regularly you can then expand this love to other parts of your body as well. Do this until you are really at peace with your body. There is incredible magic in making peace with yourself and your body.

Make friends with your body

I decided to see my body as it is: my very best friend with whom I have been spending my whole life! To be honest, there are still parts of my body I don't love yet, but I am working on that. No matter what, though, it is MY body! And I appreciate everything it is doing for me and enabling me to do it every day.

Do you want to make peace with your body? You can do that, too, and I promise you that it will probably be the most powerful decision you ever make. When you no longer see your body as a necessary evil, but as a miracle that makes your life here on this planet possible, it will change your life completely.

ENERGY EXERCISE #05 – Make peace with your body

I don't know if you have heard the saying before: "Your body is the temple of your soul." But it is so true. The body is not only there to get you from A to B, but it communicates with the world, and with us. Your body enables you to live a life, to feel emotions, see and touch things, to eat and drink, to experience so many things with your senses, to enjoy happiness, to perceive life, to live in our 3D world, and so much more.

To do this, it's important to change your focus and concentrate on what gifts your body and health are.

Take a moment to consider these powerful questions.

01 **Close your eyes for a moment and feel inside yourself.**

02 **Do you feel comfortable in yourself or are you fighting against your own body?**

03 **Do you give your body everything it needs or do you treat it like a burden that you would rather get rid of?**

04 **Do you see your body as a miracle or do you only see what you claim is missing and what should be different?**

05 **Open your eyes again and write down your thoughts about the relationship with your body: What do you think and feel about your body?**

Your body is a unique miracle

Your body is indeed the temple of your soul, and it serves you every second of your life. Your heart alone beats for you over 100,000 times a day. That's over 40 million times a year. Just for you!

And you don't need to do anything about it!

The entire physical system has been designed to maintain life for a long time. Our cells have taken on specific functions for each organ and tissue, and they have learned to cooperate with each other and to stay in constant communication. Modern medicine understands only 10 percent of what your body knows intuitively, and in reality doctors or therapists don't heal their patients. They just facilitate the body's healing system by adding whatever is lacking.

YOU are the only person that can heal yourself. Or to put it in other words, your body is its own expert when it comes to healing. Just think about the last time when you were cooking and cut yourself. (Sorry, I'm not suggesting you are a clumsy cook – but sometimes it just happens.) When you see the cut, you don't tell your body:

Please kill all bacteria

Please heal the inner skin layers

Please heal the outer skin layers

And so on

No, you put a Band-Aid on it and let your body heal the cut, right? And in 95 percent of all situations, that works fine. But what happens in the other 5 percent? From my point of view, I would say there is something that keeps your body from doing its job. But what are these blockades that keep your body from healing?

Let's have a closer look at how our body works. The human body consists of 50 trillion cells functioning perfectly under the same guiding principle, and the organ that governs and regulates that principle is the brain. The brain and the central nervous system send a constant stream of messages to all cells, creating a feedback loop of information. One side of the feedback loop runs automatically; one side is affected by our thoughts and choices. This is really important to know. Our perceptions of experiences and our day-to-day choices enter our body's feedback loop, resulting in a signal being sent from our brain to our cells. For our cells, there is no difference in the signal whether it began as an emotion or thought or came from adrenalin. The cells cannot distinguish between the two. This means that our cells are constantly listening to the signals from our thoughts and are changed by them.

We don't only think with our brains alone. All 50 trillion cells actively share your thoughts. Your thoughts can be toxic or can heal your body. Whatever they are, these thoughts turn into matter. Therefore, what we think catalyzes a change in our body. This means, you can change your physiology by your thoughts, feelings, and intentions. And that's why it is so important to create positive thoughts that support your optimal health and wellbeing.

By understanding how our body works, it becomes quite obvious that while exercising, getting proper nutrition, and avoiding toxins all play a role in overall good health, more important are the messages that your cells receive. Not only from your brain but also from your thoughts, beliefs, and emotions. Your body is an amazing mechanism, governed by the brain, and which will take care of you for life if you minimize negative messages and maximize those that are positive. Nurture every muscle, organ, and cell with positive feedback, and you will become healthier. So, start sending positive messages to your body.

Start by telling your body:

01 **I am beautiful.**

02 **I am perfect.**

03 **I am whole.**

Your entire energy field contributes to manifesting what you feel and determines the vibration with which you send information into the world. You can change this immediately with the following exercise:

ENERGY EXERCISE #06 – Yes, Yes, Yes

This is a very powerful Energy Exercise[2] That is really fun and brings you into a state of joy. Your own energy and emotional state will be raised very quickly and brought to a higher level. Happy hormones are re-

---

2 This is an exercise I learned from Laura Malina Seiler.

leased in your body, and a big grin may spread on your face during this exercise.

At first, the exercise may feel strange or unfamiliar, but that's normal. After a few times, you will notice that it becomes easier, and you will enjoy it more and more. I just love it and, honestly, you can do it wherever you are. Ok, maybe you won't feel comfortable doing it in the supermarket or in a restaurant, but hey, you can always go to the restroom, right?

01 **Stretch your arms and bring your hands above your head.**

02 **Form your hands into fists.**

03 **Then pull your firsts down with a swing so that your elbows point downwards.**

04 **While pulling down, repeat the word "Yes, Yes, Yes," many times in succession.**

05 **Then bring your fists quickly upwards over your head again.**

06 **Do the exercise very quickly one after the other.**

07 **You can do the exercise for one minute or until you feel that your emotional state has changed in a positive way.**

Why my body didn't let me down

Your body is the tool through which you experience the world, and it enables you to perceive happiness and life. Even if life is not always fun, it empowers you to do all the things you want to do.

If you want to stay healthy and balanced, it is important to take care of yourself and your body.

However, looking after your body and making sure it gets everything to function optimally is only one side of the equation, because that only takes care of your physical energy.

Looking back on my own experiences, I know that my body always aims for perfect health. But does my mind also aim for perfect health?

Understanding the power of my thoughts, my beliefs, and my emotions – some of them stored in my subconscious; some of them even stored in my energy body (more about that in Chapters Eight and Nine) – and their impact on my health and wellbeing, was a real game-changer in my life. All of a sudden, I realized that my body hadn't let me down at all! In fact, it was actually the opposite. Despite all my (negative) thoughts and beliefs, my body had always tried to function at its best. And it did send me continuous signals that I needed to change something, but unfortunately, I didn't understand these signals. Eventually, it didn't have a choice but to show me my limits, so it basically pulled the plug on me to shock me into paying attention.

Understanding this made me extremely grateful, because now I could see that my body was actually taking care of me. That it had tried its best. And landing in that deep, dark hole had offered me the chance to look at certain things in my life and react to them. That's when I realized that I could change my life every day. Every morning I can decide what I want to continue in my life and what I can change.

But I needed to apply a more holistic approach. The real secret of lifelong good health is actually taking into account ALL energy levels. Isn't this powerful? That's why we are learning more about all the other energy levels, how they influence your life, and how you can influence them. However, before we move on to these energy layers, I would like to mention one more important topic which is close to my heart. It is an incredible methodology that literally changed my whole life and the way I perceive everything around me. It is actually the most important thing I have ever learned, and I want to share it with you. I am 100 percent sure that you will love it, because everybody I have shown it to so far has been really excited about it. And I always get the response: "Why didn't anybody tell me this before?" Find out more about this incredible energy tool in the next chapter.

Your Energy Insights

We tend to focus all our attention and energy on those things that are not the way they should be or those we don't like. Are you one of these people? Well, now is the time to change it, to put all your energy on things that matter to you, and that you want to reach. The best way to start applying this approach is with your body, because there is incredible magic in making peace with yourself and your body.

The power is in appreciating what your body is doing for you all day long, and to ask yourself how you can support your body so that it can do its best job. Have you ever thought about that, about what YOU can do FOR your body? The more you align your mind and body, the more energy will be available to you, and you will feel better.

Always keep in mind that YOU are the only person that can heal yourself. Other people can support you, but you and your body are the experts when it comes to your healing.

Chapter Five

Your body wants to talk to you

POWERFUL
DECISIONS

*"We make our decisions.
And then our decisions turn
around and make us."*
Frank W. Boreham

Start communicating with your body

Have you ever thought that your life would become so much easier if you learned to communicate with your body? I actually did many times. There were many questions I wanted to ask my body, especially when it came to situations of pain or when it didn't function the way I expected it.

However, it was not until I started working with my energy that I found out that there is actually a very powerful tool to communicate with my body. This wonderful tool is the kinesiological muscle test. Do you know it? Kinesiology is a fairly young alternative treatment method, and literally means "the study of movement". The kinesiological muscle test goes back to its founder Dr. George Goodheart, an American chiropractor who, by chance in the course of his work, came across the connection between muscle tension and our thoughts, feelings, and organs.

The body reacts to positive impulses with muscle strength, and to negative impulses with muscle weakness. So, if there are energetic disturbances and imbalances in the energy flow, this is reflected as muscle weakness in the associated muscle group. This method works simply because our bodies cannot lie. And it is exactly this mechanism of the kinesiological muscle test that enables communication with your body, mind, and soul level.

Would you like to communicate with your body? Your wish is my command! Are you ready to have a conversation with your body?

ENERGY EXERCISE #07 – Discover your body's language

In order to communicate with your body, the first step is to establish the rules of communication with it. After all, when you ask questions, you need to be able to understand the answers.

So, you have to calibrate your body. The easiest way to do this is to close your eyes and stand loosely. Then say out loud: "Give me a YES!" For me, the body then tilts forward, so I know that this is my YES. But it can also be something completely different. With my daughter, for example, YES is when she tilts backwards. The first step of the communication is to find your YES.

The next step is to find out what your NO looks like. To do this, you stand and relax again. Close your eyes and say out loud: "Give me a NO!" For a NO, my body tilts backwards. But as I said, it can also stay in position, or tilt sideways. Everything is possible. In any case, it is not important what my YES and NO are, I am just sharing it so you get an idea of what it might be. The most important thing is that you know YOUR particular YES and NO!

I'll break this down in clear steps in the exercises below.

01   **Stand up. Place your feet hip-width apart.**

02   **Place your left hand on your belly button and your right hand over it.**

03   **Close your eyes and concentrate on your inner center.**

04   **Now ask the following question (loudly): "Body, please show me a clear YES." Pay attention to how your body reacts to this question and notice what you have perceived.**

05   **Now ask the following question (loudly): "Body, please show me a clear NO." Again, pay attention to how your body reacts to this question and notice what you have perceived.**

According to the movements of your body, you will now know your "YES" and your "NO".

This is vitally important to recognize for any further questions and communication!

Now that you have calibrated your body, you can start with simple questions to see if the Standing Method, and communication with your body, works.

ENERGY EXERCISE #08 – Using your body as a pendulum? – The Standing Method

At the beginning, it is important to become acquainted with the Standing Method. That's why you should start with questions for which you know the answers, to build up trust and provide you with more certainty. These questions could be for example:

01 **Is my name (your name)?**

02 **Is my name Paul?**

03 **Is it raining?**

04 **Is the sun shining?**

05 **Am I in the office?**

06 **etc.**

Now you can continue with the actual Energy Exercise as follows:

01   **Stand up. Place your feet hip-width apart.**

02   **Place your left hand on your belly button and your right hand over it.**

03   **Now close your eyes and concentrate on your inner center.**

04   **Now ask simple question like:**
               - Is the sun shining?
               - Is it evening?

05   **Pay attention to how your body reacts to this question and notice the answers.**

06   **Continue with other questions you would like to ask your subconsciousness.**

After some tests, you can ask questions where you are unsure of the answer.

Before you get started to explore the Standing Method, check out these tips:

01   Everything that concerns YOU can be tested.

02   The simpler and the shorter the question is, the better!

03   It is very important how you formulate your question. So it's good to think about what you really want to ask beforehand.

04   You have to be careful with double negations. For instance, if it's evening now and I ask the question, "Is the sun not shining?" Then a YES comes.

05   Be aware, however, that this test result is always only a snapshot and the result may look different again in a few hours.

06   The tester and the tested are united in one person. Of course, this runs the risk of "cheating" oneself during testing because your mind might step in.

07   If you are unsure about a result, check your results with counter-questions or the exclusion procedure. However, do not ask the body the same question a second time.

08   The subconscious mind immediately registers this as uncertainty and will hold up a mirror to you according to your uncertainty and give you exactly the opposite answer.

Important note: Throughout the book, many of the Energy Exercises will ask you to use the Standing Method to check in with your energy. So, you might want to bookmark this page in some way. Or if you are a rebel, you may wish to fold the top of the page over.

Questions you always wanted to ask your body but never dared

You can use the Standing Method to communicate with your body about your body, and much more. Take the opportunity to start a dialogue and exchange with your body. It is so powerful when you can really kind of "talk" with your body. Isn't this fantastic? Find below some insightful questions you always wanted to ask but never dared or knew how to get them answered:

01 **Do I take sufficient care of my body?**

02 **Do I have enough energy?**

03 **Do I drink enough water during the day?**

04 **Do I eat consciously?**

05 **Do I move and exercise enough?**

06 **Do I get enough sleep and relaxation?**

07 **Should I detox and detoxify my body again in the next two weeks?**

08 **Do I enjoy my work?**

09 **Do I have healthy and nurturing relationships for the most part?**

10 **Is my fun factor in life high enough?**

11 **Am I happy with my financial well-being?**

As I mentioned at the beginning of the book, there is *no compulsory writing*. But if you want you can write down your answers in a notebook, as you might find this helpful. Are you satisfied with the results? Is there anything you are going to change in your life? What should be different?

Do this exercise regularly, perhaps at the beginning of every week, and track how your answers change.

How the Standing Method transforms your daily life

I can give you a few examples of how I use the Standing Method in my everyday life.

I used it for correcting my daughter's math homework. One evening, she had a lot of homework, with quite a few calculations involving a lot of columns. I didn't feel like recalculating every single calculation, so I just asked:

Are all the calculations correct? NO came up.

Is more than one calculation wrong? YES.

Are more than two calculations wrong? NO. So two results were wrong.

Then I asked in which of the columns the errors were, and so within half a minute I was able to tell my daughter that two calculations were wrong, and these two errors were in the first and third columns. We had a look at the homework together and determined the two wrong calculations. Curious, I also checked the other calculations (yes, I did!) and they were all right.

So you see, using the Standing Method can save quite some time.

My daughters are using the Standing Method as well, of course. Once, when my daughter was going to bed, she suddenly realized she had forgotten to learn her French vocabulary. So, she just asked her body: Am I going to have a French test tomorrow? The answer was NO. Then she went to sleep reassured, and there was actually no test the next day. As long as you are testing questions that are related to YOU, it should work fine. But please be careful when you are testing things for other people. The answers may be biased and, hence, not correct. It is much better if you show them how they can test for themselves. That is real empowerment, isn't it?

And finally, a quite astonishing story from me. I am the chairperson of the Alumni Chapter of the Technical University Graz in Switzerland. Four times a year I organize an interesting field trip. For March 2020, I had planned a visit to CERN[3] in Geneva. A friend of mine from Germany also wanted to come and participate. In January 2020, I asked him if he was coming and he told me, "You know, I would totally love to come, but every time I ask if I should go to Switzerland and participate in the CERN excursion, I get a NO."

---

3 European Organization for Nuclear Research

This was really weird, because I knew that he had always been dreaming of visiting CERN. However, he stuck with this NO and said, "Maybe you'll do the excursion again next year, then I'll be there for sure."

I replied, "You know, the organization of the CERN visit was so complex, I won't do it again."

So, the excursion was planned for March 28th in 2020. And now we all know why my friend always received a NO answer when he asked his body if he should come to Switzerland. The visit to the CERN was, of course, canceled due to the beginning of the Coronavirus crisis.

My lesson learned from this story was that sometimes answers seem illogical, because our minds just don't see that far. But as you can see, there was a reason for the answer that none of us considered at the time. Looking back, I find this story very exciting. And, of course, I hadn't asked myself in January if the visit would be canceled, because at that time I didn't even think that was an option.

How to make strong decisions

In this context, I then thought that if I could test with the body what is good for it, shouldn't I also be able to test which decisions are good for me?

I don't know if you are aware of it, but subconsciously we make more than 20,000 decisions every day. Isn't this incredible? Most of these decisions are made in an instant – without us even noticing. It already starts in the morning when the alarm clock rings, and we decide whether to jump right out of bed or enjoy lying under the duvet for another two minutes. And it continues like this all day long.

When shopping, you almost automatically reach for the same products over and over again, but how often do you question whether these products are good for you and your family? At work, too, we find ourselves in countless situations in which we have to make decisions at lightning speed. Many of them are completely unconscious, most of them are not life-changing. But each decision needs energy.

Most people are unaware that any decision-making process can consume quite some energy, and therefore make you tired. And it makes no difference whether you like to decide quickly or think about decisions for a long time. That's why we train ourselves to love routines and be on autopilot SO much, because that's our brain's way of reducing decision-making and saving energy for the important ones.

Therefore, when you have to make big decisions, it is good to take into account that you should not make them in the evening, after a long and exhausting day, but rather in the morning when you are fresh and full of energy. And please consider the energy for making decisions in your daily overview of your Energy Balance Sheet which we spoke about in Chapter Four.

I used to have a very hard time making decisions and therefore put off many of them, no doubt because I always wanted to please everyone and not disappoint anyone. This behavior reflects in a way how the different levels of energy of the *PEEMS Model* work together. Physically our body regulates how much energy it gives to the brain to think complicated things, like an active decision. But emotionally, a people-pleasing pattern or an old belief system can still get in our way.

My mom used to sing a song about me not being able to make decisions. If we went shopping together when I was a teenager, I could often not decide for hours between two pairs of jeans, for example. I would try on one after the other, but usually neither of them matched what I had imagined. And instead of having the courage just to just say I wasn't buying either of them, I would then decide with a heavy heart on one of them. Of course, once I tried them on again at home, I always had an internal argument with myself about why I hadn't chosen the other one!

Does this sound familiar to you? How do you make your decisions on important issues? (Not that buying a pair of jeans is an important issue.) Do you make them quickly and easily based on your gut? Or do you think long and hard about them? Or are you someone who writes lists of the pros and cons of different options?

In times like these with so many changes going on, it is more important than ever to make the "right" decisions. But what is the right decision? Is there even a right decision? And if so, what does it look like? What does a good decision feel like for you?

From my point of view, there are no really right or wrong decisions. Whenever you make a decision, in that moment it is the best possible one for you. The most important thing is that you made a decision at all. So, you should be proud of yourself, because many people keep putting off decisions because they don't know exactly how and what they should decide.

This uncertainty in decisions has several reasons:

01 **That we ourselves do not know exactly what we actually want.**

02 **It is an "uncomfortable" decision, with which we may have to take a stand or cause offence.**

03 **We are afraid of hurting someone else with it.**

But if we don't decide for ourselves, then someone else – or life – makes the decision for us. That might be acceptable sometimes, but be honest, do you really want others to decide about you and your life?

For all three above-mentioned reasons, I always found it rather difficult to make decisions. Later on, as an engineer, I chose the more analytical path and wrote a list of pros and cons each time. I then evaluated and analyzed them, thought about them, and suddenly knew exactly what the right decision at that moment was.

But that is a very lengthy, and not always practical, process. That's why I was always looking for a tool that would allow me to make good and powerful decisions in an easy and quick way. Would that also be of interest to you?

Making powerful decisions with the Standing Method

Then I realized: the kinesiological muscle test is the perfect solution to making powerful decisions!

In this context, I thought that if you could connect with your body through the Standing Method, you could also use it to connect with your energy. So that's why our energy is so good at helping us to make the right decisions for us. And above all, to make decisions that are in harmony with you and your values.

So, I started making energetic, powerful decisions. To get used to it, though, it's best to start with small decisions and then move on to bigger ones when you feel more comfortable.

By having a simple tool for energy-based decisions, you can save yourself so much energy. As I've already pointed out, small decisions need only a little energy, whereas bigger decisions can consume quite a lot more. Firstly, you need less brain capacity to evaluate all the different options with all their advantages and disadvantages. And secondly, you get an answer to your question, which is good for you in the long run.

Deep in your heart you can feel when you are making the right decision. We all immediately know when we make a clear decision and stand behind it. And even if it is a rather uncomfortable decision, it is usually accepted quite quickly by others, while you feel relieved to have finally made a decision, don't you?

ENERGY EXERCISE #09 – Force or Power?
The questions for your business

Just to make sure that you are aware of it, this Energy Exercise is very powerful! Using it is not something that you should do on the side. You really need to concentrate on it, to get fully involved – with the right intention. If you only use it half-heartedly, this will reflect in your answers and they will be half-hearted as well. Then you will question what I'm telling you and decide this Energy Exercise doesn't work for you. But you would be wrong. Everybody who is ready to use it can use it and work with it. You understand what I mean? Are you ready to use it?

And this amazing exercise is the basis for all further Energy Exercises in this book, because it is so powerful and can literally change your life. For that reason, it is important to use it in the right way. Therefore, please always be aware of where and how you use it.

For example, a friend of mine, Jean, discovered that, at first, running water reversed the YES/NO response for her. "When I tested something in the shower," she explained, "the opposite came out, until I was energetically stable enough that it didn't happen anymore.

And if you have a specific question for your body, please set this intention and connect with your body *BEFORE* you start asking your questions. Especially when you are testing food or medicines, you have to be very clear that you want to talk to your body consciously.

When shopping, I find the Standing Method very practical. At the beginning, when I was standing in the fruit and vegetable section, I used the Standing Method to find out which products were good for me by asking: Are these strawberries good for me? If a YES came, I'd buy them. A NO might mean that perhaps the spray was incompatible for me, for example, so I'd leave them.

Of course, for time reasons you can't do that with all products when shopping, but with certain things I always ask – especially if you don't know exactly the ingredients or where it comes from. Nowadays, I have to admit I don't have to test anymore. When I take a zucchini in my hands, my body now reacts automatically and tilts either forwards or backwards. It is really amazing and makes grocery shopping much easier and so much more fun!

On top of using the Standing Method in your daily life, you can also utilize it to make powerful business decisions. How? It is based on the Power vs. Force in the Scale of Emotions of Dr. David Hawkins, which I shall explain in more detail in Chapter Eight. For now, just follow the process below:

01   Stand up. Place your feet hip-width apart.

02   Close your eyes.

03   Think of the question you want to ask e.g.: "Is this project power for me?"

04   Observe in which direction your body is moving – if it is a YES or NO.

05   Cross-check your answer by thinking of the same question and ask: "Is this project force for me?"

06   Observe in which direction your body is moving – if it is a YES or NO.

"Power" means that you should go for this project, because in the long run it will give you energy. Whereas, if you get "Force" as an answer, it means that doing this project will in the long run take energy from you. In other words, when you are really excited about something, that is an attractive, high frequency power that will make it much easier for you to achieve your goal. But when you are forcing something, that is a lower frequency force and makes it that much harder for you to make your idea or plan happen. It doesn't mean that you won't make it, but there is the chance that it will be more challenging for you to get it done.

The other way to make powerful decisions

With decisions it's important not only to make a choice but also to have the determination to act accordingly.

That way you will implement the decision and concentrate your energy on the success of the goal.

The Energy Exercise can help you to make decisions, but you have to implement them yourself. After all, nothing comes from nothing. If you decide to run a marathon, for example, you can't sit lazily on the sofa every evening but have to train according to a training routine. This Energy Exercise doesn't take that away from you either.

But in most cases, it is easier to stick with a decision if you know that it is good for you.

Making decisions this way is definitely a change – especially since you've been doing it differently for years. And even if the answer you get seems illogical at first glance, in the long run it might be the right answer. Sometimes it only becomes clear later why the answer was the way it was.

Your Energy Insights

Let's summarize the most important points again. If you use the Standing Method correctly, it always works. You don't need to practice for it. Or rather, the practice consists of turning off your mind. And the more often and the more relaxed you are about this, the better results you will get when answering your questions. When you start using the Standing Method, it can literally change your whole life. Just because it is so powerful, it is the cornerstone of all Energy Exercises to come.

Be experimental and enjoy it!

Chapter Six
You clean your physical body every day – How about your energy body?

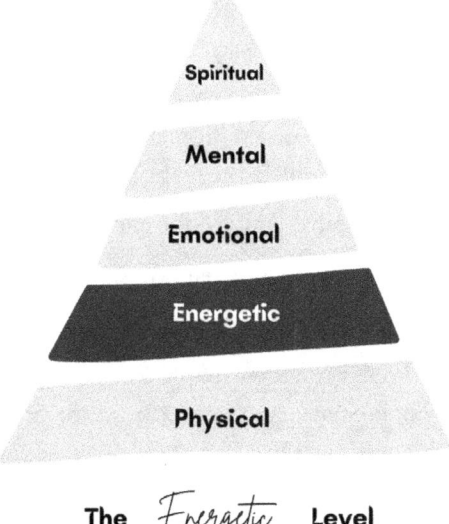

The *Energetic* Level

As you are now aware of your physical energy and know how to work with it, let me introduce you to your energetic energy.

CLEAN YOUR ENERGY

"When you realize you aren't your physical body but the energy living within it, you start to look at things differently."
Led By Source

You clean your physical body every day – How about your energy body?

Did you brush your teeth today? Take a shower? Brush your hair? I am quite sure you did. And if not today, then at least yesterday, right?

Why am I asking these weird personal questions? It is none of my business! However, when it comes to your energy, I feel a little bit responsible for it.

I am quite sure you are familiar with facts like:

2/3 of our body consists of water.

Our body consists of 206 bones.

However, very few people know that we are actually a walking battery. Were you aware of this fact? Nobel Prize winner Dr. Otto Warburg discovered that cells maintain a voltage across their membrane – comparable to the voltage of a very small battery. He found that healthy cells have a measurable voltage of minus 70-100 millivolts, with heart cells having the highest voltage (usually between 90-100 millivolts).[4]

---
4 An AA battery has 1.2 V and 9 Amp – so it is much smaller.

Every cell has an ideal cell membrane potential. If this potential is below the ideal value, fatigue, lack of concentration, and exhaustion occur. If the potential is permanently below the ideal value, the cells can no longer function perfectly, which affects the entire human system and ultimately leads to disturbed cell division. Too few, too many, or non-functional cells are formed, which leads to chronic diseases.

Dr. Warburg found that due to permanent stress — as it is ubiquitous in today's society — accompanied by a toxic environment and the natural aging process, our cellular voltage decreases. Have you noticed that when you're not feeling well, you feel drained of energy? And when you feel vibrant and healthy, you feel like you're bursting with energy? There's some science behind this. People with chronic diseases and persistent exhaustion subjectively showed decreased cellular tension of 30-50 millivolts. Cancer cells exhibit the lowest voltage at less than 15-20 millivolts. Based on these little voltages, ultra-fine direct microcurrents (normal 60-70 μA) act in cells and organs. These microcurrents are closely connected to the manifold functions of the body's own messenger substances.

Hence, every thought, every movement, every function of the body, and every healing process is coordinated by electric nerve impulses between the brain, limbs, organs, and glands. In every moment, our heartbeat, thinking, and self-regeneration, electric and electromagnetic fields play the main role. The main role in what? What is ACTUALLY going on, as opposed to what we think is happening?

I guess you know that the autonomic nervous system is mainly responsible for organizing the body. This happens **completely unconsciously**. Because it would completely overwhelm us to constantly control consciously breathing, heartbeat, blood pressure, metabolism, etc. And even more interestingly, the nervous system works de facto via a mini electrical circuit as a means of communication. Hence, our vegetative nervous system also communicates and functions energetically. And as we are 2/3 made of water, which is coincidentally nicely conductive, we are actually a *small energy circuit*. Isn't this incredible?

What has so far received little attention in conventional medicine is the fact that we are electromagnetic beings. Our bodies are made up of a complex network of electrical and chemical signals that enable us to function and interact with the world around us. The nervous system, which includes the brain, spinal cord, and nerves, uses electrical signals to transmit information throughout the body.

These electrical signals, also called nerve impulses, are created by the movement of charged ions across the cell membrane of nerve cells. The heart also generates electrical signals that can be measured with an electrocardiogram (ECG), which provides important information about the health and function of the heart. Nobody questions that a heart is restarted with electric shocks, or that pacemakers are installed which persuade the wheel to run again with a small electric impulse if it stops for a short time. The fatal part of these facts is that the electromagnetic field of most people is not in balance. This can be the real reason for symptoms and diseases.

Hence, our electric and magnetic field has a huge influence on our wellbeing. This field is also known as your energy body or aura. I am quite sure you have heard this term before. It describes the various layers of vibrating energy that surround our physical body. Most of us cannot see the energy body, but what we can all do is FEEL these energetic fields. Not only our own but also the fields of others.

As you can now see from my explanation, your aura or energy body is not so "woo-woo" or esoteric stuff, but simply biophysics. You are simply a human battery with an electromagnetic field around you, just like any other battery.

Not only your physical body is unique

And it is the same with your energy body. Also unique, it depends on your personality, the situation, your mood, and how you are feeling, what you are thinking, on which frequency you are vibrating, in which environment you are, and many more factors. And in a way this is logical. When you are sad and don't feel well, you don't stand upright, your shoulders are hunched, and you keep your head lowered. This body posture will also be noticeable in your energy body, which will shrink and be tighter to your physical body.

On the other hand, when you get some exciting news and you feel like dancing through the streets, your body posture will be upright. You will feel confident and strong, and your face will shine with joy. Your energy body will expand itself, and people around you will notice your energy and perceive it more intensely.

ENERGY EXERCISE #10 – Feel where your energy body is right now

Now that you are familiar with the concept that YOU are a kind of battery with an electromagnetic field around you that goes beyond the physical body you see in the mirror, wouldn't it be interesting to know more about YOUR energy body?

Do you know where your energy boundary is? Maybe this is your first time getting in contact with your energy body, but that's completely fine. Familiarize yourself with the idea that you consist of different layers of energy, and then let's do this together.

01   **Stand up. Place your feet hip-width apart.**

02   **Close your eyes.**

03   **Set the intention that you want to connect with your energy body.**

04   **Set the intention that you want to feel your energy body.**

05   **Ask yourself: "Is my energy body more than 1m around me?"**

06   **If you get a YES, continue by asking: "Is my energy body more than 1.5m around me?"**

07   **Continue asking, until you get a NO.**

08   **Then you can fine tune the size of your energy body by asking in the opposite direction like: "Is my energy body more than 1.4m around me?"**

09   **Continue asking until you get the actual size of your energy body.**

So, if you can feel your energy body around you, you can of course also change its size. Let's assume your energy body is 1.5 meters around you. Then you can set the intention to expand it to 2 meters around you. Visualize your energy expanding and adjusting to your intent. This size is comfortable for most situations, giving you enough room to breathe while maintaining a strong presence energetically.

Notice how comfortable you feel when you claim your space energetically!

I am quite sure you will have experienced this already at a party or an event. You are there, talking to different people, enjoying the party, and then all of a sudden somebody enters the room, and everybody notices. Most of the time this is the case when this someone has a very powerful energy body with a diameter of a couple of meters. Very charismatic people can have an energy body with a diameter of up to 20 meters, or even more.

But, of course, this also works in the other direction. Do you have someone in your family with a large energy body, and when they are in a bad mood and come into the room, everyone else's mood instantly freezes? Same. Once again, it is a powerful person, but they transport their current low frequency outside. It works this way because the energy body can not only be large, but also transports the charge of the person. Incredible how the energy of people around us can have an influence on our own energy and mood, isn't it?

A mix of energy

What I wanted to point out is that because our energy bodies surround our physical bodies, every time you stand next to somebody, the energy bodies of you and the other person are mingling and interacting. And this not only happens at home or with your colleagues at work, but also on the bus, in the supermarket, or at a concert. Your energy body doesn't distinguish between your family members and complete strangers. Crazy, isn't it?

This brings me back to my questions at the beginning. You have a routine of taking care and looking after your physical body, but how about your energy body? Do you clean it every day and take care of it as well? Probably not. Of course, you have a good excuse because you didn't know it even existed.

Maybe now you understand that because your energy body is constantly mingling with others and picking up their emotions, thoughts, or other stuff energetically, it makes sense to clean your energy body as well. And this becomes even more vital when you find yourself dealing with other people's stuff on top of your own issues!

That's why I am sharing a very simple but nevertheless purifying Energy Exercise with you. So, no more excuses!

ENERGY EXERCISE #11 – Clean your energy body

This Energy Exercise can not only be used to clean your energy body from emotions, feelings, moods, and much more that you pick up unconsciously from people around you, but also to clean out stuff you have accumulated in the course of your own life.

01 **Stand up. Place your feet hip-width apart.**

02 **Close your eyes.**

03 **Set the intention that everything that doesn't belong to you or doesn't serve you any more, leaves your body, heart and energy body.**

04 **Set the intention to collect everything that does not belong to you and send it back to where it belongs.**

05 **Everything that remains, ask the earth to take from you. Imagine how everything flows through your body, through your hips, through your legs, through your feet into the earth, where it is transformed.**

06 **Feel inside - has everything that doesn't belong to you gone?**

07 **Check with the Standing Method. If you still feel that your energy body is not clean, then go back to step 3 until it is gone.**

I do this Energy Exercise every evening. To make sure I don't forget, I have made it a habit to do it right before I go to bed. I stand next to my bed, close my eyes, and let go of everything I have picked up during the day and everything I don't want to take on anymore. Needless to say, this habit has dramatically improved my sleep.

Gone are the times when I tossed and turned in bed for hours before finally falling into a restless sleep – something I experienced for years! No more rolling over and over again the same thoughts from the past about what I should have done better and differently, worrying about my future and my family, going through my To-do lists in my mind, what I hadn't managed to do that day and absolutely must do tomorrow, and what I must not forget under any circumstances.

Cleaning your energy body and mind before going to sleep is a wonderful and liberating method to let go of everything that is bothering you and start with a new and different energy for the next day. I strongly advise you to make this a habit and do it at least once a day.

However, I should point out that you might pick up some stuff that doesn't belong to you during your dreams and notice that you don't feel as refreshed and as yourself as you should. For example, you might feel anxious even if there is no obvious reason for you to feel worried. In this case, I recommend doing the Energy Exercise first thing in the morning.

It is a little bit like brushing your teeth – once is better than never. But getting used to doing this Energy Exercise in the morning, and as a second step in the evening, provides an even better effect.

Your body has an anatomy too

The anatomy of your physical body is certainly known to you. But did you know that your energy body also has a kind of anatomy? Of course, it does! We have several hundred energy exchange centers that connect our physical body with our energy body.

There are seven main energy centers that concentrate and control the flow of energy, called chakras. The term itself comes from Sanskrit and translates to "circle" or "wheel". This means that your chakras can be seen as energy wheels which ensure that energy flows into and around your body.

Chakras function as gateways to direct energy from your energy body into your physical body via the meridian system, while at the same time releasing lower-frequency energy from your physical body into your energy body for transmutation. There is a circulation of energy that ideally flows freely without blockages.

The physical body is dependent upon this vital energy flow to maintain optimum wellbeing, but chakras can become blocked by long-held negative energies – rigid thought patterns, negative emotions, toxins, etc. When chakras are blocked or their energy flows are disturbed, there can be an undersupply in certain areas of your energy system. This in turn affects your wellbeing and can even lead to physical issues. And that is something we don't want to happen.

The seven main energy centers and what you should know about them

Each chakra has its own frequency, is responsible for the energy in a particular part of the body where it fulfills individual tasks, and is associated with a certain color. However, each chakra can change its color in the course of the day depending on your situation and how you are feeling. So it is ok if your chakra doesn't always have the color it is supposed to have.

01 **The Root Chakra is located at the pubic bone.**
It stands for grounding, stability and I trust life.
Color is Red.

02 **The Sacral Chakra is located a hand below the belly button.**
It stands for partnership, joy of life and I am creative.
Color is Orange.

03 **The Solar Plexus Chakra is located a hand above the belly button.**
It stands for self-worth, decision making and I create my life.
Color is Yellow.

04 **The Heart Chakra is located in the middle of the chest.**
It stands for love, compassion and I am love.
Color is Green.

05 **The Throat Chakra is located in the middle of your throat.**
It stands for communication, self-expression and I am creative.
Color is Light-Blue.

06 **The Third Eye Chakra is located in the middle of your forehead.**
It stands for wisdom, conciousness and I express myself.
Color is Indigo.

07 **The Crown Chakra is located in the middle of your crown.**
It stands for connection to the Universe and I am a part of everything and everything is a part of me.
Color is Lilac.

Before we jump into this exercise, I would like to share some information about a very important chakra with you.

What most people don't know – The power of the Womb Chakra

Even though I have been managing my personal energy for quite a while now, there is one thing I only discovered recently[5] – the power of the Womb Chakra. Have you ever heard about it? The Womb Chakra is said to be the most powerful energetic point within the body, but unbelievably I had never heard about it before. There are stories that because it is so powerful, this information was only passed on in special circles. So, I was curious to find out more, and to share my findings with you.

Thanks to gender equality, all human beings – women as well as men – possess a Womb Chakra. Why? The explanation is that as we each enter this world through our mother's Womb Chakra, even men without a physical womb have the energetic imprint of the Womb Chakra and access to its power. In its pure original state, the Womb Chakra represents the entire power of creation, pure love, and absolute clarity. It is also seen as the divine feminine energy within each of us.

By checking with the Standing Method, I found out that although everybody has a Womb Chakra, it seems to be deactivated in most people. But when I activated my Womb Chakra, I have to admit that nothing special happened. I found this confusing, because I had expected to feel a difference afterwards, yet everything stayed the same. So, how was this powerful and magic?

However, I changed my mind about the Womb Chakra once I activated it for one of my clients, Randia, during a session. As I already knew from my own experience, at first she didn't feel a difference. However, two

---

5 From Stefanie Bruns

days later she called me and was completely enthusiastic. She explained – which I could hardly understand at that time – that she felt like being in a very powerful tornado. She felt calm and fully energized at the same time. Everything seemed to be swirling around her at high speed. She had a bulk of energy, like an energy surge. She explained it like this: "Activating my Womb Chakra carried me to another level of energy. There were so many things changing in my life. Three weeks ago, I was completely unemployed. Now I am teaching in a master's program about blockchain and crypto."

ENERGY EXERCISE #12 – Activate your Womb Chakra

After Randia's experience, I realized that activating the Womb Chakra is not enough. For some people it seems to also be necessary to unblock or cleanse it before it can operate optimally. So, I checked with the Standing Method that my Womb Chakra was activated but needed some cleaning. Then I cleaned it according to the process described in Energy Exercise #9. Immediately I could feel the difference. It was not such a surge of energy as Randia had described, but more the feeling of all of a sudden having a warm sun shining in my belly. It is such a beautiful and comforting feeling, and I am very grateful that my Womb Chakra is now activated AND operating properly.

As you can see, there are differing experiences when you are activating your Womb Chakra, but there is no right or wrong. Don't put too much pressure on yourself; just experience how it evolves. It will be right for you.

With an activated and working Womb Chakra, we experience a peaceful inner landscape and strengthen our ability to walk in the world knowing our truth. Want to try to activate your own Womb Chakra? Let's try it out!

01 **Close your eyes and connect with your Womb Chakra by thinking of it.**

02 **Ask yourself: "Is my Womb Chakra activated?"**

03 **If you get a NO, set the very strong intention to activate it.**

04 **Check with the Standing Method if it is activated. If not, repeat the process.**

05 **Continue by asking: "Is it working optimally?"**

06 **If you get a NO, check with the Standing Method whether you need to clean it.**

07 **Clean the Womb Chakra according to the process described in Energy Exercise #9.**

08 **Check with the Standing Method if your Womb Chakra is clean and optimal.**

09 **Connect with your body and assess if you feel any difference.**

If you still don't feel any difference in your energy, there is the possibility that your Womb Chakra is blocked. I will show you how you can unblock this in Chapter Eleven.

ENERGY EXERCISE #13 – Chakra tapping and Recharging Method

To keep your chakras functioning (and this is important for your health and wellbeing), I'd like to share a straightforward activation technique with you.

01 **Stand up and close your eyes. Tap your Root Chakra and say (to yourself or loud):** "I bless you and fill you up to 100% with red energy."

02 **Tap your Womb Chakra and say (to yourself or loud):** "I bless you and fill you up to 100% with turquoise energy."

03 **Tap your Sakral Chakra and say (to yourself or out loud):** "I bless you and I fill you up to 100% with orange energy."

04 **Tap your Solar Plexus Chakra and say (to yourself or out loud):** "I bless you and I fill you up to 100% with yellow energy."

05 **Tap your Heart Chakra and say (to yourself or out loud):** "I bless you and I fill you up to 100% with green energy."

06 **Tap your Throat Chakra and say (to yourself or out loud):** "I bless you and I fill you up to 100% with blue energy."

07 **Tap your Third-Eye Chakra and say (to yourself or out loud):** "I bless you and I fill you up to 100% with violet energy-"

08 **Tap your Crown Chakra and say (to yourself or out loud):** "I bless you and I fill you up to 100% with white energy."

When using the Chakra Tapping and Recharging Method, you will notice that your body will tilt forward when you start tapping each chakra, and your body will tilt back into the upright position when the chakra is balanced and filled up. You can do this Energy Exercise whenever you get out of bed, and you will be amazed at how much energy you have to start the day.

Your Energy Insights

The physical body consists of energy that vibrates very slowly, which is why it appears to our physical eyes to be solid. This has led us to focus mainly on the physical aspect of ourselves without recognizing that it is actually created from and sustained by the energy body that is outside of our normal perception.

You are quite aware now that you are a walking battery with an electro-magnetic field around you. So, depending on your personality, the situation, your mood and how you are feeling, what you are thinking, on which frequency you are vibrating, in which environment you are, and many more factors, your energy body can be the size of your body (meaning very small) or the size of a city (meaning very big).

By working on your chakras and your energy body, you can expand your energy body. And you can also shrink it. Experiment with what suits you best in different situations. And yes, everybody can do this – you, too.

YOU are an energy being, so it is natural for you to manage your personal energy! You just kind of forgot about it. Now it is the time to reactivate these skills!

Chapter Seven

Immediately recharge your energy without a magic wand

RECHARGE YOUR ENERGY

"The higher your energy level, the more efficient your body. The more efficient your body, the better you feel and the more you will use your talent to produce outstanding results."
Anthony Robbins

Immediately recharge your energy without a magic wand

Even if you are not in a dark hole, we all know these moments in our lives: after a long and exhausting day, when the children are finally in bed, we would actually like to do something we enjoy, but we feel much too tired, we lack the energy to get up, so we just keep sitting on the couch falling asleep while watching Netflix. Have you experienced that as well?

When I started working with my energy, I was looking for a tool with which I could recharge my energy in an instant. I wanted to have something at hand to recharge my energy in the moment when I needed it, not only to get out of my dark hole, but also because I wanted to be there for my family, for my little girls. I needed energy when I was with them, but also when I went back to work in my demanding job.

I longed for a fairy with a magic wand who could come by and charge me up with energy in a flash whenever I needed it. Wouldn't that be fantastic?

Here comes the disclaimer: I am still not a fairy. But I'd like to show you how you can do this yourself – without a magic wand.

The key is the access to the quantum field

The key to recharge your energy is the quantum field energy. And the great news is that energy is available in abundance in the quantum field! There is an unlimited supply for everybody!

As an engineer, I know that – according to the law of conservation of energy – energy is neither created nor destroyed but always only transformed from one form into another. By looking closer into this, I realized that there is unlimited energy in the quantum field. I just needed to get access to it. But how should I do that?

When I was with my daughters, I observed that they had much more energy than me. No doubt when you think back, as a child you probably also had an infinite amount of energy. Remember all the things we had to learn? For example, you kept trying to stand up – and fall down – again and again, until you finally managed to do it. This is because you always automatically charged yourself with the energy from the quantum field. In the course of our lives, however, we have kind of unlearned this direct charging mechanism. As a result, we have cut ourselves off from this natural recharging of our energy body, and therefore we sometimes feel tired, lacking in energy, and exhausted.

Of course, there can be other reasons for being low on energy, including growing responsibilities and different stresses due to our busy lifestyles. This means that we have a high utilization of our mental energy which leads to the situation that up to 60 percent of our energy can be used by our brain and autonomous nervous system. In contrast to that, children are playing much more, so they use more of their physical energy. While children still have a lot of play in everything, many adults are on autopilot and often too tired to recharge their batteries. They don't even know that they can actually manage their energy at all.

But it does not have to be this way. The possibility of recharging ourselves with this unlimited energy from the quantum field is always there. We have only lost access to it, and with this simple method I would like to show you how you can find this access again and charge yourself with energy in a quick and practical way.

ENERGY EXERCISE #14 – Recharge your personal energy

Before we start, feel inside yourself for a moment. Are you reading this in the morning (and a morning person) and are still full of energy, or in the evening, feeling tired and stressed at the end of a long, exhausting day? It is important and valuable to develop a feeling for where you are standing energetically at the moment.

Once you are connected with your energy, move on to experience the following Energy Exercise:

01   **Stand up. Place your feet hip-width apart.**

02   **Place your palm of one of your hands upwards.**

03   **Close your eyes.**

04   **Then think or say aloud: "Please charge me with divine energy."**

05   **Imagine that a powerful golden light beam is flowing in your palm and from there through your arm and your whole body.**

06   **Test with the Standing Method if your energy level is now up to 100%.**

07   **If not, repeat the whole process.**

Sometimes you don't feel the energy right away. It takes a couple of minutes until you realize that you feel less tired and have more energy. It is a process; sometimes it needs time. But you will get there. Be open to it and observe.

Cheeky disclaimer: If you want to super-charge, it is perfectly fine to use both hands!

This Energy Exercise works on three of the five energy levels of the PEEMS-Model, namely the physical, energetic, and mental levels. These energy levels can be refilled up to 100 percent, especially your energy body, and all your chakras can be easily recharged with this Energy Exercise. It is a little bit like filling up the fuel tank of your car.

You might be thinking this is great news. And with the possibility to recharge my physical energy with direct energy, is there even a need to eat and sleep? That's a wonderful thought, isn't it? I hate to disappoint you, but unfortunately, some physical limits of your body cannot be removed... yet. Your body is still your body. On the physical level, a certain battery charge definitely goes with energy alone. But other parts will continue to go only with action, like eating, drinking, exercising, and sleeping. So, there will still be moments when your body shows you that charging energy alone is simply not enough – similar to your mental energy, where you still need to take a break or rest.

However, this Energy Exercise allows you to recover much faster, or to need fewer or shorter breaks. And you can even recharge your physical energy to have more energy to do more exercise. You are unique and so is your body, so you have to experiment with what works best for YOU and YOUR body.

In the very beginning, I did this exercise several times a day. To make sure I didn't forget to do it, I linked it to something I automatically do every day: brushing my teeth. I brush my teeth three times a day and have gotten into the habit of always recharging myself with energy right afterwards. That way, I don't forget. And after recharging in the morning, I start the day with a lot of energy and find that everything is much, much easier!

When I notice during the day, for example, that my energy level is decreasing or I'm getting a bit tired after a few hours in the office or after a strenuous meeting, I treat myself to this Energy Exercise instead of a coffee or a snack. Of course, you don't want your colleagues to watch you doing it, but it is a quick and easy exercise to do in the restroom. Yes, you read right: for the restroom. My colleagues must have wondered more than once what I was doing in there, as I always came out of the restroom with a big smile and full of energy.

The beauty of it is, the more often you do the recharging exercise, the less your energy level drops and the less often you need to do it. And I can tell you that it is a wonderful feeling to go through the day with so much energy and to still feel I could move a mountain in the evening. You can extend that feeling even more when you combine this exercise by grounding yourself.

You don't need a lightning road, but grounding works wonders

Have you ever heard of grounding? The key focus of grounding yourself is to focus on the here and now. But don't worry, you don't need to install a lightning rod at your body to ground yourself. On the contrary, it is all about connecting more with your body, because a lot of stress and anxiety comes from disconnecting from our bodies. When you are connected with your body, you can simply learn to shift gears more consciously, to feel in your body when it overreacts, then counteract it. The more grounded you are in your body, the less stress and anxiety you will experience, because it gives you the chance to take an honest look at what your body is actually trying to tell you.

Grounding is just the direct connection with nature and earth. Especially in summer, I am a big fan of walking barefoot as often as possible. It makes such a difference to feel the ground with the soles of your feet. There is nothing more empowering than walking with your bare feet on dewy grass. The grass is moistening and tickling my feet at the same time, and this makes me feel connected with nature and life. I just love these moments just for me.

Take off your shoes and walk at the beach or in the woods. Feel the pebbles, grass, forest floor, or sand under your feet. Of course, there should be no hurtful elements. Walk on the natural ground for ten minutes, consciously breathing in and out slowly. Focus on the texture of the ground. Direct skin contact with the irregular surface of the earth relaxes you. You will experience a natural massage of the soles of your feet, and you will be able to concentrate on the here and now. You feel peaceful and relaxed. You experience nature and natural grounding. For city folks, it has the same grounding effect when you take care of your plants (if you have any) in your apartment. Tasks including watering, cleaning, cutting off withered parts, etc., help you find balance and enhance your wellbeing.

However, most of the time we are not out in nature to experience this "pure" grounding. That's why I developed the following grounding exercise which will allow you to ground yourself wherever you are and whenever you need it. You can use this Energy Exercise to become calmer and more grounded in various situations, like when you have the feeling that you have lost contact with your body, before a call with a new client or an important meeting, as well as in difficult or stressful moments. It can also help you to get a clearer picture of what is bothering you after you have grounded yourself.

ENERGY EXERCISE #15 – How to ground yourself

The first time I introduced this to my client Rena, it was amazing to see the shift of energy before and after the exercise. It is hard to describe, but she felt powerful and calm at the same time. That gave her the feeling of being very strong. The next time she used the Energy Exercise was before she had a very important meeting. She was very anxious about this meeting but went to the restroom beforehand to ground herself, and she achieved amazing results. Feeling so strong gave her so much self-confidence that she could make her point and present her new project idea to the management. And guess what – of course, it was very well received and accepted.

Do you want to feel that strong yourself? Then please try the following Energy Exercise:

01	**Find a comfortable position either sitting or standing.**
02	**Take a long deep breath, and then slowly breathe out. Do this a few times.**
03	**Now, envision that energy light roots are growing out of your feet and find their way to the center of Mother Earth.**
04	**Let the red and powerful red energy from Mother Earth flow up through your roots to your feet right into your body.**
05	**Take in Mother Earth's powerful and comforting red energy and radiate it throughout your body and all your cells.**
06	**Now envision that a light column beams up from the tip of your head through the whole universe to meet the Central Sun.**
07	**Let the bright white sparkling energy flow down through your energy column through your head into your body.**
08	**Take in the sparkling and inspiring white energy from the Universe and radiate it throughout your body and all your cells.**
09	**Try to visualize the energy flow between Mother Earth and the Central Sun and how the energies mingle to a beautiful and warm rose energy.**
10	**You are now standing strong as a pillar of light between heaven and earth, fully grounded!**

This exercise is very useful when you want to feel strong and powerful – like a tree which is deeply and firmly rooted to Earth. With this exercise you have the foundation with which you can anchor and connect yourself energetically to the core of the Earth.

Being grounded, you can align yourself to your core, which will make you much more peaceful and balanced. You might even start viewing your situation from a different perspective and be able to solve problems faster for the greater good of all involved.

Bovis? What the hell is Bovis?

When you have read this far, you will probably remember that I mentioned Bovis twice already. I hope you are curious, as I will now reveal the secret behind it. As mentioned in Chapter Two, there was one thing that puzzled me as an electrical engineer when I started working with my personal energy. Why was there no unit to measure personal energy? We cannot see and hear electrical energy, but we can measure it in kWh. Hence, I was convinced that if it was possible to measure electrical energy, then it must also be possible to measure personal energy. As an engineer, I wanted to make the progress of my energy work tangible. Not only for me, but for everybody.

It took me a while but finally I discovered it. There is actually a unit with which you can measure life energy, and it is called Bovis. It owes its name to the French radiesthesist Andre Bovis (1871-1947), whose goal was to develop a simple quality control for food in the 1930s. With this method, he wanted to quickly and easily measure the freshness and vitality of food, to have an indicator of the beneficial value of food for the human body. Interesting, isn't it? Even though I am an electrical engineer, I hadn't heard of Bovis until rather recently!

Based on his observations and experiments, Andre Bovis developed a scale with values between 1 and 10,000 Bovis. Based on his scale, he measured the individual Bovis values with a pendulum or a tensor. He found that a value of 7,000 Bovis should be the minimum for health and food, and that's where he drew the line. The main points of the Bovis scale are as follows:

01   The lower the value of an object on the Bovis scale, the worse it is for you.

02   Everything that has more than 7,000 Bovis can have a supportive effect on your organism.

03   Everything below 7,000 Bovis takes energy away from your body.

04   Food and substances below 3,000 Bovis are even extremely detrimental.

These are the values of the original Bovis scale, but now the scale is set to open upwards.

Right from the beginning, I was completely fascinated by Bovis. And from the moment I started using the Standing Method to evaluate my personal Bovis values, my life changed completely. I am now much more mindful of what I eat, where the food comes from, and if it gives my body energy. I also learned which activities give me energy, which is extremely helpful to know when I need a boost of energy. I also adapted my night and morning rituals to make sure I start the day with lots of energy and high Bovis values. If you want to do the same, it is really simple!

Do you know how to measure your own Bovis values?

Using the Standing Method, let's measure how much personal energy you have in Bovis. The most important part about knowing your Bovis value is not that you have the exact number of your own energy, but to know what kind of food and environment is good for you. By measuring the Bovis values of the food you are eating, then comparing it to your own Bovis value, you learn whether it is giving you energy or taking energy from you.

If you nurture your body with food that has more Bovis units than your body (e.g., more than 7,000 Bovis), then you are supplying your body with uplifting energy. Such foods empower you. Therefore, it is important to be able to determine for yourself which food supplies your body with vital energy and which don't. It is really fun and brings so much insight into your life!

ENERGY EXERCISE #16 – Measure your Bovis values

01 Stand up. Place your feet hip-width apart.

02 Close your eyes.

03 Ask (out loud): "How many Bovis do I have?"

04 Ask (out loud): "Is the value above 5,000?" YES?

05 Ask (out loud): "Is the value above 6,000?" YES?

06 Continue to ask this question until you get a NO.

07 The number before you get a NO is your personal Bovis value.

When you are testing your Bovis value for the first time, you can fine tune it by reducing the steps. Let's say you get a NO at 9,000 Bovis. Then you can continue to ask: Is the value above 8,500 Bovis? If you get a YES, you can continue from there until you get your next NO. Let's say you get it at 8,700. Then you can continue to ask: Is the value above 8,650? If you get a YES, you can narrow it down from there.

At the first attempt, it might take a little bit longer to determine your Bovis value. However, you will get used to the process and you will become faster and faster. You will find that you soon start with your number with the first YES, then your body will remain in this position while you count, and shift in the NO position when you reach the value. So, what is your Bovis value right now? I just love experimenting with Bovis.

Measure your Bovis value (in the morning) before you use the Charging Method. Take a notebook or journal and write it down.

Measure your Bovis value after you use the Charging Method of Energy Exercise #13. What is it now? Write that down in your notebook or journal as well.

ENERGY EXERCISE #17 – Measure if food is suitable for you

Being able to test whether something is good for you or not can massively simplify your life. I recently read a book by Anthony Williams in which he described freshly squeezed celery juice as one of the healthiest foods on earth. Have you heard of it? Since I always like to integrate new bio hacks into my life, I immediately bought a juicer, and every morning I made fresh celery juice and drank it. After a while I noticed that I always got a stomachache after drinking the celery juice.

Eventually – late, I know – I tested whether the celery juice was good for me and my body, and the answer was relatively clear: NO. (If only I had done this before I bought a juicer). But as you can see, I sometimes forget to use the Standing Method, too – especially when my mind takes over. And, as an engineer, that is often the case.

You can use the Standing Method to clarify whether certain foods are compatible with you or not, to decide whether the product is good for you.

If you want to find out if food is suitable for you, you can test it with the following Energy Exercise:

01   Stand up. Place your feet hip-width apart.

02   Close your eyes. Think of the food you are holding in your hands or in front of you.

03   Ask (out loud): "Does this food have more Bovis values than me? Does it give me energy?"

04   When you get a YES - go ahead and enjoy your food.

05   If you get a NO - you can change its energy by saying: "I bless this food with divine energy."

06   Then check again: "Does this food have more Bovis values than me? Does it give me energy?"

07   If you get a YES - enjoy your food.
If you still get a NO - I wouldn't chance it.

There are always tricks on how you can increase the Bovis values of food or drinks. A friend of mine, Irma, just loves drinking coffee. When we tested her and found that coffee was taking energy from her and actually harming her health, she was devastated. So, we looked into different possibilities of how we could increase the Bovis value of her coffee. After several experiments, we finally found the solution – and it was as simple as almond milk! It turned out that by adding a little bit of almond milk to her coffee, we increased the Bovis values in a way that it became more tolerable for her body. I cannot tell you how happy she was about this solution.

You now have a powerful Energy Exercise on hand to check out if the food you are eating is actually good for your body and contributing to your health. Think about it for a minute – this is amazing! But always be aware that you are testing what is good for YOU. We are all unique (see my story above), so what might be good for others won't necessarily be good for you, and vice versa. Please keep this in mind.

"Kraftorte" and their Bovis values

With Bovis you cannot only measure the energy of food, but also of your environment. There are so many things you can actually measure with it. Healthy trees have Bovis values above 10,000, which explains why we feel better after a walk in nature. The healthier the forest, the more we can recharge there. This is also true in mountains and at the water or seaside. In nature, we can recharge our batteries easily and efficiently.

Likewise, certain places have particularly high Bovis values. These so-called "Kraftorte" or power places have been used for ceremonies and rituals since ancient times. The Bovis value is increased in such places because the radiation of the earth is especially concentrated, so they are found in the immediate vicinity of springs, in grottos, and in caves. For example, the Rhine Falls in Schaffhausen show 17,500 Bovis.

Especially sacred buildings, which were erected in the pre-Christian times as places of worship, can have extremely high Bovis values. Strong, sacred places and churches, such as the Grossmunster in Zurich or the Cathedral in Chartres, have 18,000 Bovis. The Taj Mahal is one of the top energetic places in the world, with 750,000 Bovis. How many Bovis do you have at your place? Check it out easily with the Standing Method, and please let me know – I am curious!

ADVANCED ENERGY EXERCISE #18 – Find your optimal Bovis values

With the Standing Method, you also have the possibility to test the optimal Bovis values for yourself, because they are individual. For example, you can test your optimal Bovis value for food/drinks and the optimal Bovis value for living space (your apartment or house). My optimal Bovis values for food are 12,000 Bovis, and for my living or office space both 20,000. With the following Energy Exercise, you can find out yours as well.

In the Energy Exercise I show the process of finding the optimal Bovis value for your food. To find the optimal value for your living space, you can use the same process.

01 **Stand up. Place your feet hip-width apart.**

02 **Close your eyes.**

03 **Ask (out loud): "What is my optimal Bovis value for my food?"**

04 **Ask (out loud): "Is the value above 10,000?" YES?**

05 **Ask (out loud): "Is the value above 11,000?" YES?**

06 **Continue to ask this question until you get a NO.**

07 **The number before you get a NO is your optimal Bovis value for your food.**

So, what does this number mean to you? You should always increase the Bovis values of your food, as described in Energy Exercise #17, to your optimal value. Food with that much energy really serves your body at its best. If you try it out, you will experience that your food tastes better. Food is then no longer just food. It is actually POWER FOOD.

Your Energy Insights

If you really want to manage your personal energy, you need to measure your progress – at least at the beginning. So, it is important to track your development.

This can easily be done by determining your Bovis values, for example, before the Recharging or Grounding Energy Exercise, and afterwards. You can measure the transformation in your energy immediately. You don't only FEEL it, but also TRACK it. Isn't it such a cool tool?

Knowing how to work with Bovis also broadens your scope. All of a sudden you are more aware of what you are eating and drinking and, I guess, also of your environment. There are so many possibilities to experiment with it, and it is just so much fun! You can even do it as an activity at your next party. You will find out that your guests just love it!

Chapter Eight
Make friends with your emotions

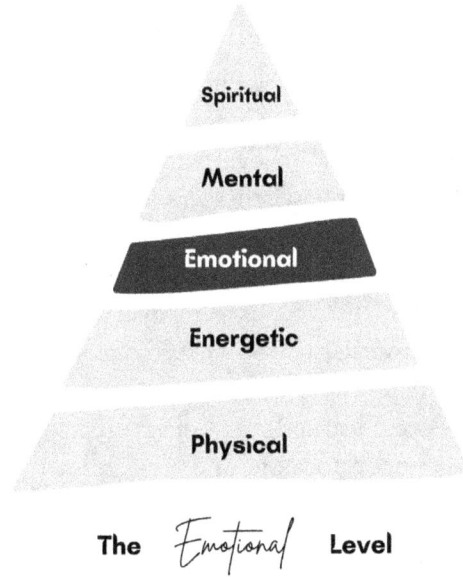

After our deep dive into your energetic energy, we now move on to work on your emotional energy. Old and stored emotions reside in our energy body. And because they have such an impact on our behavior and how we perceive life, they even have their own level within our energy body. If you are brave enough to face your emotions, let's have a look!

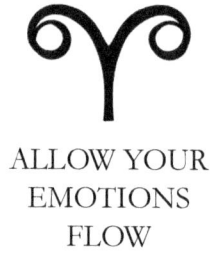

ALLOW YOUR
EMOTIONS
FLOW

"Don't hold on to anger, hurt or pain. They steal your energy and keep you from love."
Leo Buscaglia

Make friends with your emotions

In order to move into the energy world, your feelings are key. Are you scared of your feelings? Join the club! I never learned how to deal with emotions. When I look back at the time when I was a child, I was taught and experienced that emotions are not something you show. It was even seen as a weakness, because it makes you vulnerable and you have to deal with the consequences afterwards. If you show your emotions, you give up your power – that's what I was told and observed. Hence, I learned very quickly that emotions are something uncomfortable and unnecessary that you better push aside.

I remember very well one extremely traumatic situation when I was six years old. I was in terrible pain, and my mom had to rush me to the hospital for an appendectomy. I was so scared. I didn't know what was going to happen; nobody explained it to me. Furthermore, at that time parents were not allowed to stay in the hospital. So, when I woke up after the surgery, nobody was around. I felt completely lonely and left behind. My father taught me to be strong and not to show any fear or other emotions, so I had to deal with my anxiety, sadness, and all the other feelings, completely alone. Since I didn't know how to get rid of them, I decided to ignore them and push them aside. I encapsulated them all somewhere deep in my body and hoped that they would never show up again.

Do you have a sealed capsule of emotions somewhere in your system? Unfortunately, I think we don't only have one of these toxic capsules, but quite a few. Do you agree?

Back then, I thought that these emotions showed up out of the blue. If they came from somewhere, they would go back there as well, so why bother about them? However, decades later, my body showed me that encapsulating these emotions and storing them in the body was actually not the best idea. It made me learn the hard way that all the emotions I had pushed aside (and there were a lot) didn't go away. They actually stay until the day you choose to deal with them.

Years later, I noticed that my digestion wasn't functioning as it should. I had always been grateful for having a healthy digestion, because I knew that not everyone does. Then all of a sudden, things changed, and I didn't know why. I hadn't altered anything in my diet or lifestyle. For me, everything seemed to be the same – except my digestion.

I guess you already know what the cause of my problem was. And now I know too, but back then it took me ages to understand that my stored emotions were causing the problems with my digestion. Now I understand that emotions always want to be acknowledged and seen. The amazing thing is, once I cleared out these emotions and let them go, my digestion started to work fine again... as if nothing had ever happened. Crazy, isn't it?

Before I explain how I managed to clear my emotions and how I let them go, let me share some background information with you. Understanding why emotions are there in the first place and how to deal with them makes the whole process easier and applicable to you. Are you ready to go?

Emotions are not just feelings, but the meaning we assign a given situation

Emotions are reactions that you experience in response to an event or a situation. The type of emotion you experience is determined by the circumstance that triggers the emotion. Hence, emotions have a strong influence on our daily lives. We make decisions based on whether we are happy, angry, sad, bored, or frustrated. We choose activities and actions based on the emotions they incite.

Eight basic emotions are universal throughout human cultures:

01	**fear**	*05*	**joy**
02	**disgust**	*06*	**happiness**
03	**anger**	*07*	**trust**
04	**sadness**	*08*	**anticipation**

These basic emotions can be combined or mixed together, in much the way an artist mixes primary colors to create other colors. We have all felt angry, shy, scared, or embarrassed at some point in our lives. Having emotions is a universal experience.

But just because everybody is having them, it doesn't mean that they are easy to deal with. I'm sure you agree that emotions are complex. Let's bring some light into how to perceive an emotion. There are four components of feeling an emotion, and this four-step situation-attention-appraisal-response sequence is called the Modal Model of Emotions.

01 **The situation you are in (whatever is happening to you at that moment).**

02 **The details you pay attention to.**

03 **Your appraisal of what the situation means for you personally.**

04 **Your response, including the facial and physical changes (like blushing or shaking), and your behaviors (like shouting or crying).**

Understanding your emotions, why you are experiencing them, and where they are coming from, can help you to navigate through your feelings to live with greater ease and stability. Of course, you can protect yourself from allowing yourself to feel anything. But that generates a subscription to future legacy and problems, and in my view it means missing out on so many positive feelings as well.

How to deal with your emotions

Our emotions are not the problem. We are just conditioned to respond to strong emotions by running away from them. Accepting emotions as they come helps you to get more comfortable with them. To practice accepting emotions, try thinking of them as messengers. They are not "good" or "bad"; they are neutral. Maybe they bring up unpleasant feelings sometimes, but they are still giving you important information that you can use. Ask yourself: "What are these feelings telling me?"

In a way, emotions are following the concept of coming out of the blue and going back there as well, as I experienced as a child. I just didn't understand the main part of the concept, which means watching them flow through your body without hanging onto them. At six years old I felt overwhelmed by all these emotions, and that's why I hung onto them. Now I know that I should do the opposite. I need to show them, feel them, and then let them go.

Let me explain what I mean. When I started writing this book, I had a call with Cassandra to discuss the outline of the first chapter. Of course, this particular chapter was already in my head, and I knew exactly what I wanted to write about. But when I started talking to her about it, I found myself becoming very emotional. I felt a huge sadness coming up, and some tears began rolling down my cheeks. I actually had no words, and I was struggling with myself, but I realized that this was a part of my story that hadn't shown itself so far and I needed to work on it.

I sensed that Cassandra felt a little bit uncomfortable with me starting to cry (most of us don't feel comfortable if somebody around us cries), so she offered to stop there and continue our discussion another time. She could feel that it was very intense and emotional for me. However, I knew that I just needed a couple of minutes to settle the situation, so we agreed to pause for five minutes.

When I came back to the call, I was full of energy. Any sadness and tears were gone, and I felt so much clearer, lighter, and full of energy. Cassandra was a little bit startled at the complete change in me and asked me what I had done. So, I explained to her that this was the whole concept of Personal Energy Management, and that I had just released some huge emotions. Once the emotions were out of my system, they were no longer a burden. Some people might have been scared by this and thought they couldn't handle it at that moment, but I was actually grateful for it. I understand how energy management works. Emotion is energy in motion; and it needs to come out. Then it's no longer a burden to me.

After my explanation, Cassandra was even more excited to work on the book. She said: "You have already helped me today by understanding this process. I cannot wait to read about this Energy Exercise and do it for myself."

Already, in just our first session, I had helped Cassandra to understand her emotions better, and that made my day. It is the whole reason why I have written this book. As humans, we have so many misconceptions. We worry about tears and emotions and feelings as though they are negative. But they are the opposite.

I am always very grateful when these emotions show up. I had stored them, because for some reason they were too big to handle at that moment and I couldn't process them. Now they were showing themselves and coming out. When this happens, you need to know that this rush of emotions is only temporary.

So, you need to step aside from them and just allow them. Showing them is freedom; freeing yourself and your body from old and stored emotions. I have learned how to deal with them in a different way, and that is the beauty of Personal Energy Management.

The Hawkins Scale of Emotions

Another little science warning! Cornelia the scientist is stepping in again with something intriguing to share with you.

Energetic biology states that people vibrate at different frequencies, depending on what affects their personal energy. Based on these findings, Dr. David R. Hawkins, a physician and psychiatrist, developed the Scale of Emotions. Do you remember in Chapter Two when I talked about how we are all vibrating on different frequencies, and that every thought and emotion has its own frequency? I know that chapter was a little "dry" to read, but for all further energy work it is so important that you understand that in a nutshell you are born with your very special unique core frequency. However, due to life, your experiences, traumas, and things you were taught, your unique core frequency has changed – as you can see from this very simple equation:

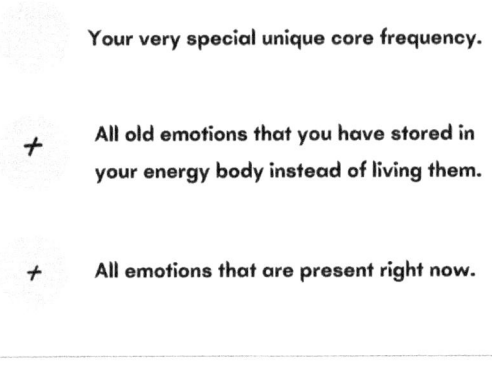

What a discovery! Emotions have a frequency that co-vibrates with your unique core frequency. The Scale of Emotions, which you can see in the next picture, explains the different vibrations of emotions and how they can impact you.

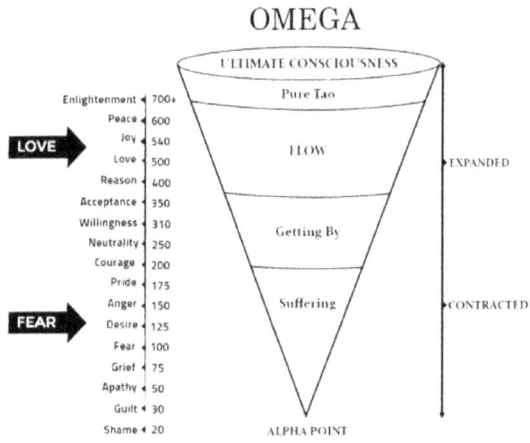

Source: Map of Consciousness – allesistenergie.net

In the figure, you can see that emotions such as shame or guilt are at the lowest end of the scale and thus belong to the lowest vibrating emotions. In contrast, positive emotions such as love and joy are emotions that vibrate at a much higher frequency. According to Dr. Hawkins, the lower the vibration you vibrate at, the more energy you consume. In contrast, you add energy to the world when you are vibrating at a higher level. It is important to understand that no emotion or level is good or bad *per se*.

Because emotions have a higher or lower frequency, and thus give versus take energy, the emotions as such are never bad. Despite what many of us have been taught, any emotion always has value.

Emotions are nothing more than lights on the dashboard that indicate your condition. You would never say that the orange fuel light on the dashboard in your car is bad or stupid, right? It just shows you that you should refuel. And it is the same with your emotions. Many unpleasant emotions actually show you that something doesn't fit underneath, that there is something you SHOULD LOOK AT instead of constantly ignoring it.

Most people seem to vary throughout their lives as to what emotion, and therefore frequency, they are operating from. Unfortunately, most are vibrating below 200. However, everybody will have a "base level" which is the regular frequency they are normally vibrating on. This scale also helps you to understand why certain measures with which you raise your vibration are only temporary. I am quite sure you have already experienced this during and after a meditation. After a wonderful meditation we all feel calm, loved, or powerful. Unfortunately, on most occasions this state doesn't last long. Lower frequencies always have a higher intensity than the high ones. The reason for this is that as soon as you are back in your "normal" life, you fall back on the frequency you are normally vibrating on. The emotions of your "baseline" are so much stronger, so if you have a lot of fear, anger, or sadness in you, these feelings always prevail in the long run. For this reason, doubt, fear, and guilt must be replaced by higher frequencies of love, courage, and joy.

How does this work?

That's why it is so important not only to raise your vibrations short-term, but to clear out the cause of these low-vibrating stored emotions and the emotions themselves in the first place. Do you see the point I am making? Raising your vibrations is very important to experience and help you in many situations, but it is only a Band-Aid. It is not dealing with the stored emotions and why they are stockpiled in your body. This is the beauty of Personal Energy Management. Not only does it deal with acute emotions in your life, but even more importantly it gives you Energy Exercises to clear your body from all these captured emotions and set them free again. Want to know how to do this? You will be startled by how easy it is.

The 90-Seconds Rule

Before we look into how to liberate your stored emotions, from my experience I think it is necessary to understand what an emotion actually is. STOP! Before you skip this section because you are fed up with your feelings, I really urge you to keep reading. In general, it is always better to have some more insights about what is actually bothering you, because then it is easier to deal with it, right?

Generally speaking, emotions are something quite natural. This means that we cannot prevent ourselves from experiencing and perceiving the whole range of emotions again and again. So sometimes we cannot influence our feelings and emotions; they just show up. However, what you can influence is the way in which you perceive and deal with them.

My approach is that every emotion is in general a feeling in motion. It is something in motion. So normally it is supposed to flow through you, and the longest will be 90 seconds. Yes, you read correctly – just 90 SECONDS.

This means that an emotion, such as anger, lasts only 90 seconds from the moment it is triggered until it naturally disappears. If it lasts longer – which, by the way, is the rule for most of us – it is only because we unconsciously make sure that we "prolong" this emotion, keeping it alive longer than would be naturally (biochemically) necessary.

This is often the case if something happens or is too big for us, or too humiliating. And it is particularly the case if, as a child, we never really learned how to deal with (big) emotions. Then we just hang onto this emotion and it's kind of stuck in our system.

There might also be the possibility that you don't want to feel that emotion, so try to run away from it. But doing that will end up keeping that emotion actively high, and alive longer, and longer, and longer. Or we prolong the stressful feeling because now, based on the first feeling (e.g., I'm angry), we are mentally engaging with that anger even more. We then go even deeper into our anger, and thereby pour even more oil on the fire. Or you get angry with yourself because you feel angry. There are many reasons why we decide to hang onto an emotion and, hence, prolong it staying with us.

However, if we get it out in 90 seconds, we are all good. If you take yourself back in the role of an observer and say to yourself, "I allow myself to feel angry or sad. It's okay, you're allowed to be here," then it just flows through. There may be another emotion comes afterwards, but then you just keep observing and letting go.

Most of the time, we hang onto our emotions, especially as kids, because we never learned how to deal with them. But these emotions are then stuck in our system, and this can sometimes have an impact on our physical body if they are stored for a long time.

The easiest thing is to support emotions in their natural rhythm, meaning: let them come and go. It's a little bit like sneezing. This is something I have made a habit. Whenever an emotion is coming up, either because of a thought, a situation I tapped into, or somebody has said something to me, I now take the time to go fully into it. And I find it really easy, because now I know that an emotion never lasts longer than 90 seconds, if – and this is the big IF – I don't hang onto it and just watch as a kind of observer.

Basically, emotions are a little bit like small children. They want your attention. They want to be seen and acknowledged. When you look at them, and they are allowed to tell you what they want to tell you, they then flow through you or dissolve into the air and go back where they have come from.

ENERGY EXERCISE #19 – How to apply the 90-Seconds Rule

What does this mean in practice? Whenever I feel an emotion coming – the most prominent ones are fear, anger, and sadness – I literally try to take a step back and follow this process:

01 You acknowledge that you are feeling this emotion.

02 Yes, you even go a step further and allow yourself to have this emotion, by saying to yourself: "I am really feeling angry right now."

03 Allow yourself to have good reasons to feel angry by admitting: "Yes, I'm annoyed right now. My body shows me that a border was crossed and I defend myself." Or: "I did not get what I wanted and now I am stubbornly angry." But don't analyze to death.

04 Then wait 90 seconds and observe how the anger floats through your body.

After that, you will observe that you feel less angry – that the emotion has just gone.

I'm sure you are thinking, is it really that easy? Are you telling me that just waiting 90 seconds as an observer would have saved me from so much anger, pain, sadness, and guilt? The simple answer is: YES. But of course, we are all human beings, and that's why sometimes we want to feel a certain emotion longer, because we think it is part of the process.

Mila once said to me, "You know there was this person I was dating. After some time, he realized that we were not the perfect fit and he decided to break up. It hurt me a lot, and at that time I didn't let go of my sadness. On the contrary, I enjoyed being sad about him, not wanting to be together at that moment. I also enjoyed missing him, because it showed me that I deeply cared and that I am sad about it. And there was this fear that I would not find someone else and then I would be alone. But now it is ok to let it go."

I want to stress that you don't have to hang onto your emotions. Of course, you can if you want to spend some time grieving about a loss or experiencing some pain because you think you deserve it. That is ok. But I just want to let you know that you don't HAVE to.

ENERGY EXERCISE #20 – Let go of your emotions

As I have already explained, there are many powerful emotions we deal with in the course of our days and lives. You can use the Energy Exercise I am showing you here for *any emotion* that shows up. By consciously allowing and feeling your emotion, you can release and transform it.

01 Stand up. Place your feet hip-width apart.

02 Close your eyes.

03 Ask yourself: "Am I still carrying an old emotion that is draining my energy?"

04 Focus on the emotion in your body, paying attention to where exactly you feel it.

05 You don't necessarily need to know which emotion it is, but of course you can check with the Standing Method by asking: "Is this emotion x - y - z?"

06 Breathe intensely into this part of your body for 90 seconds.

07 Allow yourself to fully feel the emotion. As long as you ignore the emotion and don't want to look at it, it will remain stored in your body.

08 While feeling this emotion try to face it without judgment. Just notice it.

09 Once it is fully there, set the intention to let it go.

10 Release the emotion by imagining the emotion is flowing out of your body.

11 Check with the Standing Method if this emotion is gone.

In most cases, not only one emotion but several emotions are stored in the same place. With the Standing Method you can check if there is more than one. In that case, you ask ALL emotions to show themselves and let them go at the same time. If this doesn't work, you just do the Energy Exercise several times until all emotions are released. You will get there! And the more you practice, the easier and faster it gets. It is all a matter of practice.

Another important point is that most of the time you don't need to KNOW which emotion it is. Sometimes you will FEEL it, and it will feel familiar to you. You will know that this is anger, hate, or sadness. But don't get yourself carried away that you need to KNOW which one it is. Sometimes emotions are stored which are difficult to name because they are not so well known. So, please keep in mind that it is more important to LET IT GO than to know what it is.

The key is to approach your emotions from a position of strength. There, where we look, can dissolve, but where we look away, that will remain. The stronger your trust in yourself becomes to deal with your emotions, the easier it will be.

Why stored emotions are an energy vampire

By reading this chapter, hopefully you can understand why stored emotions are such an energy vampire. It is not that you are carrying around all these stored emotions. It is not that your whole system needs to deal with these emotions which consume a lot of energy. No, the worst thing is that they are energy blockades and keep your energy from flowing freely, and that has a vast impact on your wellbeing.

When you look at that Scale of Emotions, you see that shame and guilt are the two emotions vibrating the lowest. Interestingly enough, fear and anger are low vibrating emotions but not the lowest. With the Scale of Emotions, it is easy to understand from an energetic point of view why guilt and shame are the biggest energy vampires. And guilt can be two-fold, whether we blame ourselves for something we did to others or we blame others for what they did to us.

Why? Because when we feel guilty or ashamed, most of the time it is due to something we did or said which we wished had never happened. We wish we could erase that moment or situation, that we could wipe it from our memory. But it is still there, even if you hide it deep in your system and you don't want to be reminded of it because you feel so bad about what you did or what somebody did to you.

I would like to share the experience of my client Mila with you. When we spoke about feeling guilty, she told me a story of when she was a teenager. She was with some friends, and in the course of the evening they started teasing one of her group. Mila was very into it and pushed things further and further until all of a sudden it was not funny anymore… for any of her friends. She realized that she had stepped over a line, and she had hurt him, even if it was not intentional. Even two decades after that event, she still feels ashamed about what she did. Tears were streaming down Mila's cheeks as she told me, and I could feel how much she regretted what she had done.

We unlocked all these stored emotions and finally let them go. On top of that, she also asked her friend energetically for forgiveness. (You will get to learn this in Chapter Ten, so please don't miss out on it. It is so powerful!) When Mila finally let go, I could immediately see the difference. She stopped crying, and she looked calm and relieved at the same time. When I asked her how she was feeling, she responded immediately, "So much lighter! It feels as if a huge stone just fell off my heart! This is incredible!" These are the magic moments when I see how energy management can transform lives. Within 15 minutes your world can look completely different. It makes me feel humble any time I am part of such a magical moment.

By the way, you can use the same process when you blame another person for something they did to you. When you have an accusation against someone else, regardless of whether what the person said or did was right, these thoughts keep your body at a low frequency. And frustratingly, you are actually harming yourself the most, while the other person has probably already forgotten about the incident.

Don't want to be controlled by your emotions any longer? Just let them flow

Sometimes you are having a thought about somebody, or you hear something which triggers a memory and some emotion comes up. Normally we are brought up to suppress these feelings. We just put our emotions in a capsule and lock them away. And for a long time, I did the same.

Before working with energy, there were times when such a rush of emotions might have consumed me or taken over for the rest of the day, the week, or even the rest of the month. I wouldn't have known how to deal with that, so my biggest fear was that once emotions come they stay for a long time or even take over. Being aware of your stored emotions and knowing how to deal with them is very powerful.

Now I know and just let them flow. Now I out myself; sometimes I even cry in the supermarket. And the crazy thing about it is I don't even feel embarrassed about it. Because if something comes up now, I know it's over in 90 seconds. The interesting thing about it is that most of the time people around me ask, "Are you alright? What is going on?" And I simply answer, "It's all right. I'm just processing some emotions."

It is a little bit like if you need to sneeze in the supermarket, right? If you have the urge to sneeze, you just sneeze, don't you? But we have never really learned how to support somebody who is crying. We feel helpless and don't know how to comfort other people. For the same reason, most of the time we prefer to cry alone (in a dark room). We don't want to embarrass anybody, and we don't want to be embarrassed.

But how about if you reframe the whole crying thing? If you see that this emotion is just sweeping through your body and the tears are cleansing it and flushing everything out? And if you are honest with yourself, in most cases you feel so much calmer and relieved when you let your emotions flow and get them out of your body.

So, my advice is, let them out. It only takes five minutes, and then you can congratulate yourself: Another emotion out of your system! It's not trapped anymore. It is not blocking anything. It's not going to turn into anything. It's gone.

Emotions don't want to be in your system at all, you know; that's not their task. They just want to be felt, then go along and flow out of your body. They definitely don't want to be a blockage in your system and cause any trouble. So why not change your perspective? Imagine that you are an emotion that just wants to flow through your body, and all of a sudden you are captured and imprisoned in a dark space. Would you want to be there? No.

So just let your emotions out. Let them flow and go. It will make such a difference in your life.

Feel and flow, then let them go!

The view behind the scenes

What I really enjoy now is that I know how to deal with my emotions. I am not at their mercy anymore or need to hide them. Instead, they help me to look behind the scenes and, in certain situations, find out why I am experiencing them in the first place. When something is triggering me now, I take a moment to check where it is actually coming from. Why is the emotion there? Then I can dig deeper to unravel the real cause underneath and let the stuck emotion go. What a relief! What a transformation to discover the real me below all this stuff! I don't need to become somebody else; I just need to discover my real me!!

It is such a relief to know how to deal with my emotions, let them go, and heal the cause of why they were in my life in the first place. Most importantly, I now know how to deal with the energy blockades that kept me from living the life I dreamt of as a kid. Now I am removing one layer after the other that I put on during my life, to find my true me!

Here I would like to put in another big disclaimer: Please be nice to yourself in this process.

This is the way to raise your own frequency sustainably.

How to raise your vibration

Now you know what an emotion is and how it feels. It brings you a glimpse behind the scenes and is nothing more than a message, and if allowed to deliver that message, it will only last 90 seconds. This is a meaningful way to deal with any emotions in the present. If you know and learn how to deal with emotions now, you can prevent many new emotions from coming in and being stored. And that is a big first step!

However, as you probably already experienced and have read in this chapter, not only are there emotions passing by right now. Unfortunately, we all have many, many stored emotions in our energy bodies. The next step – which can be done in parallel – is to put in the work to get all the old emotions out of you, by doing some spring cleaning. Be aware, though, cleaning up the old and stored emotions can take months. You could compare it to the time it takes to properly clean up the basement and attic of your house or the cellar of your apartment.

The beautiful side effect of letting go of all these energy blockades is that I now have so much more energy! It is just amazing. Gone are the times when I woke up exhausted and even more tired than when I went to bed the night before. Gone are the times when I didn't have any energy for doing things in the evening like meeting with friends, going to the gym, reading an inspiring book, or just enjoying some me-time.

By learning to manage your emotions, you not only improve how you communicate with others – and hence your relationships – but you are also able to get off the "emotional rollercoaster", even out extremes in mood, and bring your life into balance. As you learn to tolerate even unpleasant emotions, you'll discover that your capacity for experiencing positive emotions has grown and intensified.

As you develop the capacity to recognize and better understand your own emotions, you'll find it easier to appreciate how others are feeling, and this improves how you communicate, which helps your personal and professional relationships to flourish.

Your Energy Insights

Negative emotions like sadness, anger, and fear can sometimes challenge us, but I firmly believe that they bring us important messages. Instead of suppressing them, give them space and ask yourself: What is this emotion trying to tell me?

Emotions serve as a catalyst for change and can help you recognize your limits and steer your life in the right direction. If you allow yourself to accept them and work with them, you can use their power to transform them. They are part of your path.

Allow yourself to live your emotions, let them come, be there, then go again. The more you agree and process your emotions, the more you are actually able to feel positive emotions. Your thoughts, emotions, and beliefs are all just energy, and if you know how to manage them you will have much more energy for living your life the way you want.

The most important thing for your mind to understand is: If you feel angry, or sad, or whatever, this feeling is NOT going to stay if you don't hold onto it; if you just allow yourself to feel it and let it go. It is ok that you feel it. You now have a powerful Energy Exercise to deal with your emotions.

And always be aware of the fact that you are *experiencing* emotions – you are not your emotions. Whatever you are having, you can also let go.

Chapter Nine
Clean up your thought garden

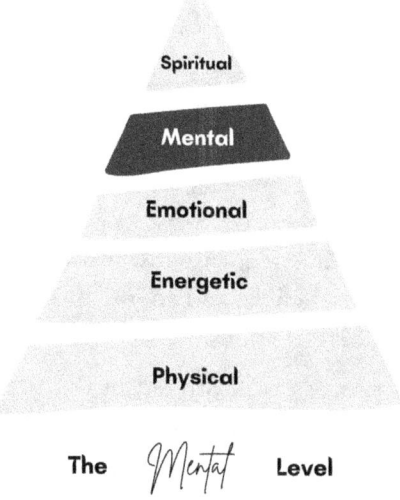

The *Mental* **Level**

When experiencing an abundance of mental energy, you will feel efficient, focused, concentrated, and motivated to deal with tasks. However, unfortunately in the busyness of modern society, you might relate more to the feeling of having a lack of mental energy. You may have felt like being on the verge of burnout, found yourself procrastinating, and that you just didn't have the capacity to take on any more tasks. Take yourself back to a situation where you were handling lots of tasks, work was hard, and home life was emotionally exhausting. That is the feeling of a lack of mental energy.

CLEAR YOUR BELIEFS

"Energy flows where attention goes."
James Redfield

Where does your energy flow?

There are so many things we need to deal with at the same time, consciously and subconsciously. It is amazing how many balls we are juggling at times, so during the course of the day it is important to observe where your attention is focused. Most of the time we are not aware where our energy is flowing to, because automatic programs, beliefs, and energy blockades are consuming a lot of our energy.

This process is illustrated in the Identity Circle below. It shows that everything with a thought comes from your belief system. This thought leads to an emotion, resulting in an action. This action leads to a certain experience, and all these experiences add up to your identity. But this is only who you think you are! Your thoughts and beliefs mirror your identity and create your reality.

Identity Circle

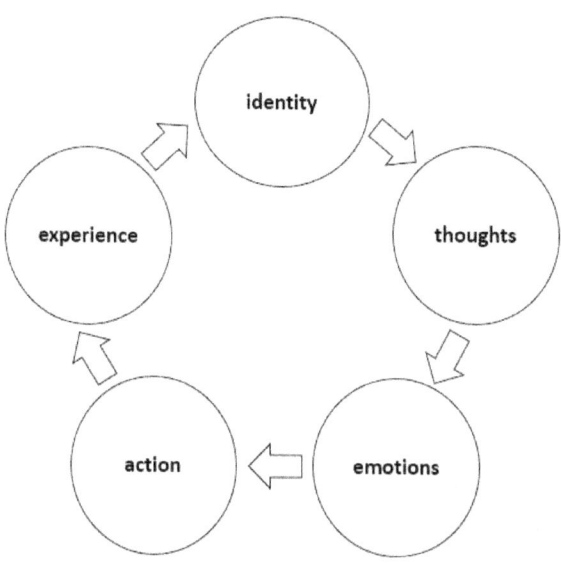

You always experience what you actually *believe*. Your external world is always a mirror of your inner world. When you look around and contemplate your life, you know your thoughts and how you think about yourself.

However, as already pointed out in Chapter Two, from an energy perspective you are born with your own special and unique core frequency. That is the real you! However, in the course of your life a great deal of different information has been modulated on your very special frequency. By that I mean every thought, every emotion, every belief, every event; everything we carry around with us has a frequency. The frequency you are vibrating at depends on your core beliefs, what you subconsciously believe in, what you think, and how you feel.

I promised to show you a powerful way to clear out these frequencies that are having such a big impact on the life you are living. The bad news is, though, there is not only one frequency that needs to be removed. Unfortunately, in the course of our lives we have collected quite a bunch of them. So, the whole process is a little bit like peeling an onion. Every emotion you let go, every limiting belief you clear out, every event from the past you heal, every person you forgive, brings you a step closer to your real you. It is a process, but once you get started you will feel the difference almost immediately. And you can work on different areas of your life in parallel. You can work on everything you become aware of, let it go, and clear it out. It's like creating a brand-new room for new thoughts, beliefs, and experiences to come into your life! Doesn't that sound great?

And the good news is that every step brings you back to the real you, to the unique core frequency you were born with.

Do you really believe this?

Your beliefs are always present. Limiting beliefs are opinions and thoughts that you personally believe to be the absolute truth, but they can also stop you from moving forward and growing in your personal and professional life. They are typically formed during childhood through the things we see, experience, and learn from our parents and from other significant people and role models in our lives.

Limiting beliefs cause your brain to filter all of your experiences to back them up. Our consciousness can only deal with a small part of all the information we absorb every day. The rest is taken in by the unconscious mind, which filters, generalizes, and stores experiences according to the values and beliefs that you hold.

Therefore, limiting beliefs not only have an impact on what you do in life, but also on how you see the world and others. This is also why limiting beliefs are sometimes called self-fulfilling prophecies. Believing a certain thing about yourself, your mind is orientated by that belief and will seek out situations or filter the experience to prove it's true.

At any given time, you are trying to support and reinforce your current belief system. This means that no matter what your inner worldview looks like at the moment, you are always trying to confirm it through external events in your life. If you believe that life is bad and unfair, then that is EXACTLY what you will look for and perceive.

If you believe that you are worthless and others can trample on you all they want, then you will constantly look for events that confirm that thought.

If you believe that you are always in the wrong line at the supermarket (the longer one), then that's what you'll get.

And what happens when, for once, you get in the shorter line? Simple: you will NOT notice. You always notice the queue only when it corresponds to your inner belief system (which is: I am always in the longer one).

As a result, you can NEVER grow beyond your inner belief system.

Clean up your thought garden

Your inner worldview determines WHAT you perceive and at the same time HOW the outer world looks to you. And THAT is exactly the reason why so many people are stuck for so many years and rarely get ahead...

Deep down they nurture beliefs that are 180 degrees opposite to their world of wishes.

And then those wishes cannot become anything.

Buddha said it already:

"We are what we think.

Everything that we are, arises from our thoughts.

With our thoughts we create the world."

In this sense, start TODAY to clean up your thought garden and remove limiting beliefs from your system. It is unbelievable how it works. Lena and I were in an international working group together ten years ago, then she moved to England with her family, and we lost track of each other. When I was looking through my contacts on LinkedIn before Christmas last year, I came across an old message from her, so

I spontaneously asked her if she would like to meet virtually to exchange ideas.

When I then saw Lena in the Zoom call, I must admit I was a bit scared (she probably was, too). The last few years had left their mark. She immediately told me that she had already had a burnout and was now on the verge of the next one. She felt drained, exhausted, and could hardly sleep anymore. During our conversation, the reason for her burnout quickly became clear. She simply could not stop working. She constantly felt the urge to sit at the computer and felt bad when she wasn't working.

Her family and those around her were naturally concerned and told her: "You need to work less!"

"You need to look after yourself more!"

"Take a break!"

But she couldn't. And yes, you already know why. The cause behind her constant need to work was a very strong belief!

We all have our limiting beliefs, but each and every one of us has different ones and to varying degrees of intensity. This depends on the environment in which we grew up. One of the best-known beliefs, which often leads to burnout, is: I have to do everything perfectly!

Other common beliefs are:

- I have to please everyone!
- I have to make an effort!

- I have to be strong!

Lena's belief was: I have to work more!

And that's why she just couldn't stop, even though she had already worked so much.

Together, we took a look behind the scenes to find the cause and where this belief had come from. We found that as a small child Lena had only received attention and recognition from her father if she continuously stuck to one thing. This motivation had become stronger and stronger over the course of her life and had taken over to the point that it was now endangering her health and wellbeing.

In the next step we resolved the cause and also the belief. Finally, we replaced this with appropriate and benevolent new beliefs, such as "I enjoy life"; "I am loved the way I am". When we were done with this, a completely new person sat across from me in the Zoom call. Lena's facial features were completely relaxed, and she told me right away that she felt like she looked ten years younger.

Her whole energy had changed, and I could clearly feel that all the pressure and tension had gone. When we said goodbye, she said to me, "Do you know what I'm going to do now? I'm going to close my laptop and go for a walk. I can't remember the last time I did that."

Two weeks later she wrote to me: "My energy level has risen and continues to rise. I feel much better and have lost two kilos, which I have wanted to do for a long time." And that was over Christmas, when there is always the danger of *gaining* two kilos!

For me, it is constantly amazing that within half an hour you can shift a situation from which a person has suffered for years, simply by using Personal Energy Management. And if you are curious now how this works, let me show you.

The main limiting beliefs

This is a list of the main limiting beliefs which almost everyone has, or has previously experienced. These beliefs may block your development:

01 I am not enough, not worthy.

02 I am a bad, wrong or deficient person.

03 I am not lovable.

04 I am not important.

05 I am not smart enough/I am too stupid.

06 I am too short/tall/fat/thin.

07 I am not creative/athletic/musical enough.

08 I am always afraid.

09 No one will listen to me or what I have to say.

10 Bad things always happen to me.

11 Other people hold me back from my own potential.

12 I am not good with money.

13 Relationships only cause pain.

14 I am not worthy of being loved.

15 You have to work hard in order to be successful.

16 I can never do what I really want to.

Of course, there are many more limiting beliefs. Who am I to have this, be this, do this?... No-one will ever take me seriously... I'm never going to succeed, so why even begin?... I don't deserve... The list is unlimited, but these examples hopefully raise your awareness of your OWN limiting beliefs.

Do any of them resonate with you? If you are unsure, you can easily check with the Standing Method.

In order to proceed, you can also sit down, close your eyes, and ask yourself which limiting beliefs you are having. Then wait for what comes into your mind and write down your OWN list of limiting beliefs. You will be astonished at what comes up, but please don't judge yourself – just write it all down. Then afterwards, check with the Standing Method whether you are *really having* this particular limiting belief.

I like to work with the question or statement, "I still have limiting beliefs that are holding me back from my highest potential." Or also with: "There are limiting beliefs that I may let go of now."

And please always keep in mind: once you are aware of this belief, you can work on it. So, it is good that you know what it is.

Some inspirations for supportive beliefs

You can always replace limiting beliefs with positive and supporting ones, like the examples below. It is better you adapt the beliefs you want to activate suitable to the situation you are working on. Try one on and watch the magic spark.

01	It's ok to be myself.	09	I am loved.
02	It's safe to shine.	10	Life happens FOR me.
03	It's easy for me to learn new things.	11	The world is a friendly place filled with love and compassion.
04	I have time to play in business and in my life.	12	I am more than enough.
05	I always have more money than what I need.	13	What I do is valuable.
06	I have more success than I dream of.	14	I reach my goals with ease.
07	It's easy to achieve my dream lifestyle.	15	I am worthy.
08	I love money and money loves me.	16	I love and respect myself.

Beliefs, like words, have power. What you tell yourself in your own head has an immense influence on your self-image and self-confidence. The most powerful supporting belief is: "I matter!" Use these supportive beliefs as a guide to find the ones that are important for you and are suitable for the situation you are working on. They help you to start thinking and feeling better about yourself to boost your self-esteem.

ENERGY EXERCISE #21 – Eliminate limiting beliefs and activate supporting ones

As I mentioned before, it is important to release and eliminate limiting beliefs and replace them with positive and supporting beliefs. Before you get started, you need to have at least one limiting belief you want to work on. Don't worry, that isn't usually a problem. Everybody has at least a thousand limiting beliefs, so this is a good point to start working on them.

Choose one limiting belief that stands out to you from your list and start working on it according to the following steps:

01   **Stand up and close your eyes.**

02   **Check with the Standing Method whether you are having one of the limiting beliefs from the previous page.**

03   **If you get a YES, check with the Standing Method if you either need to**
- let it go
- eliminate it
- heal it
- or cancel it

04   **If, for example, you need to let go this belief, set the intention and say: "I now let go this belief."**

05   **Check with the Standing Method if this limiting belief has gone.**

06   **Ask if you should activate a supportive belief instead of the limiting belief.**

07   **If YES, choose a supportive belief (from the list) and activate it by saying: "I activate this supportive belief."**

08   **Check with the Standing Method if the positive belief is activated.**

When setting an intention during the Standing Method, your body reacts to this intention. Normally, it tilts forward when you set the intention, and moves back to the upright position once it has been implemented. The amazing thing is that you can release as many, and as quickly, as you want.

When you test that you have a particular limiting belief, it means that it wants to be released. It is ready to go or be shifted. Just get started with the most important ones and then keep going. There might come a point when you have released all the limiting beliefs that have shown themselves. But if you continue managing and transforming your energy, maybe other limiting beliefs will show up. So, it is always worth checking regularly if there are other limiting beliefs there which you can work on.

I know this sounds like you are probably never done, but don't get discouraged by this thought. As I've already pointed out: One less limiting belief to carry around with you is one less limiting frequency on top of your unique core frequency. We are onions, and there are many layers to peel off. But the journey is always worth it. If you look at your Personal Energy Management from this perspective, it will keep you motivated enough to keep going.

Transforming your limiting beliefs can create big changes in your life in a very short time. So, I would encourage you to keep pursuing this journey to overcome your limiting beliefs and start living the life you desire.

Who do you blame?

Another important energy blocker is blame or accusation. As long as we hold onto blame and accusations, we give another person power over our own wellbeing. We pass on responsibility for ourselves, and with it, our power to choose how we feel. We take a victim's point of view and often later become abusers ourselves when – because of our own hurt – we hurt others.

As long as we hold blame for life, for our parents, for our partner, or even for ourselves, something will always hold us in the past. We give up our own power, and it is difficult for us to feel free and move on. Don't get me wrong, there may well have been things that happened in your life that were not right, and somebody did or said something to you that hurt you a lot. However, carrying on something that happened and cannot be changed anymore doesn't hurt the other person at all; it only hurts *you*. You are the one walking around with this pain. You are carrying this blame.

By letting go of it, you allow this person and what happened to no longer have any further impact on your life!

You free yourself!

You get your power back!

You shift your energy!

This can be so liberating, as I have found from my own experience. Just recently I discovered one of my big blames I had held onto against my wonderful mom. When I was visiting her, all of a sudden I got a terrible sore throat – I could hardly swallow. By looking into the reason for having a sore throat, I remembered that it was something I had suffered from frequently when I was younger, and which only got better when I moved out. Now with all my energy tools, I decided to have a closer look at the reason behind the pain in my throat.

Pretty soon I realized that the reason behind it was this incredibly big blame against my mom. She is a very special person and I love her very much, but when I was little, she was working very hard all the time with my dad in their business. There was little time for playing, hugging, and enjoying each other's company. I could see the little Cornelia standing there, all sad and just wanting to be hugged and loved by her mom. So, I connected with this younger me, took her on my lap, and spoke with her. I gave her all the love and attention she needed, and finally we could agree to let go of this blame against my mother. The blame that she never showed her love to me the way I needed it at that time.

Afterwards, we forgave my mom because she had done the best she could do then. She didn't show me her love the way I wanted it, with playing, hugging, and laughing, because she didn't know that this was important to me. And she had never experienced that from her parents when she was a child. Of course, she showed her love for me in her own way – but I couldn't see it at that time. Knowing all this and being a mom now myself, I knew there was no reason to hang onto this blame. On the contrary, I wanted to free myself from this because it wasn't doing me any good.

When I cleared out everything, I felt so relieved and so much lighter. Needless to say, my throat stopped hurting right afterwards, and my relationship with my mom improved. There is a deeper connection between us now, and I can show her much more how I admire and love her.

ENERGY EXERCISE #22 – Let go of blame and accusations

When you start dealing with your own blame and accusations, ask yourself the following powerful questions:

01 **Who would I be without this blame?**

02 **What would be possible through it?**

03 **What wisdom could I find in this experience, if I no longer resist having had this experience?**

Then start with the Energy Exercise and let go of your blame as follows:

01   Stand up and close your eyes.

02   Ask yourself: "Do I have any blame? Am I blaming myself or anybody else?"

03   When you get a YES, decide consciously that you want to let go of that blame. That you don't want to carry it on and let it influence your future life.

04   Observe how your body is tilting forward when you set the intention and backwards when you release the blame.

05   Ask yourself: "Did I let go of this blame?"

06   Ask yourself: "Is everything cleared out?"

07   Observe how your energy has changed.

Blame and accusations are often connected with forgiving yourself or others. So, forgive yourself for holding onto this blame for so long. This can be combined with the exercises in Chapter Ten, in which we will be diving into forgiveness. Keep reading. It will be worth the wait.

It is very powerful to get your power back!

Never underestimate your belief system

Never underestimate the power of your belief system. It has so many automatic programs running, and we are not even aware of most of them. However, our belief system is the filter on how we see ourselves and the world around us. It is the key to our wellbeing. Stefanie Bruns developed a very interesting formula, and I found it quite powerful to see it from this perspective:

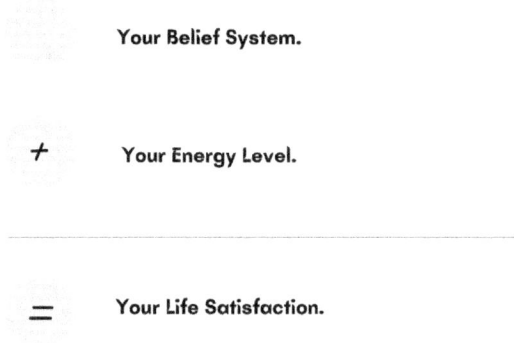

What does this formula mean?

In short, it means that your belief system influences your energy level and, hence, how happy you are about your life. In practical terms, it means that when you clear out limiting beliefs you increase your energy and raise your vibration, which leads to a higher quality of life. You feel more content and satisfied with yourself and with your life. And you feel the difference when your beliefs are in line with your values and what you want to achieve. Fascinating, isn't it?

You might argue that sometimes you're not even aware of what's going on in your mind. And I agree that I don't always have time to ponder whether I am anxious (or happy) or not. But your body knows!

How many times have you felt anxious or worried about something and tried to think yourself out of the situation? But your head went round in spirals? Anxiety crept in, and you couldn't think yourself out of the situation? Like fireworks going off again and again?

The reason for this vicious circle is that if your head got you into a situation, your head can't get you out of it. You can't use the same energy to fix a situation that has caused the situation. It just makes things worse, doesn't it? You could sit for an hour and try and think, for example, about feeling worthless and you still don't know what to do. Right?

One possibility to get out of such a situation would be to treat yourself with an energy healing session to work directly on the energy of the situation. You will notice that this makes you feel better, and there's nothing wrong with that. And you can do this on your own. I'm now giving you the Energy Exercises that you can use immediately, wherever you are. And these Energy Exercises can help you get yourself out of situations on a different level.

The charm of Personal Energy Management is that you can work with it right away and on yourself. For some reason we all think that shifting your mindset takes time. But this approach is not true when you shift your beliefs by shifting their energy.

This means you have the power to leave beliefs behind you and to create balance, because with Personal Energy Management you are working on a different level of the problem. That's why it is easier to solve. You are the creator of how you solve a situation. With the Energy Exercises throughout this book, you now have more options. It is up to you if and how you use them!

Your Energy Insights

You know what the real magic is about Personal Energy Management? The secret is that shifting your energy is so much easier than shifting your mindset. Working on your mindset is all about overwriting your beliefs and trying to become somebody else. Energy work is all about letting go. You don't need to become somebody else – you just need to let go of all the frequencies that do not belong to you.

And it is so much easier to let go than to become somebody else.

Chapter Ten

Before the sun sets, forgive

FORGIVE
AND LET GO

"Whatever energy you put out there, that's the energy you are going to get back."
Elie Tahari

Before the sun sets, forgive

The key to a peaceful life is all about forgiving. Many people have trouble forgiving, because what happened or what another person did to them was not right. But in a way, you are the one who is dealing with that situation and accusation. Probably the other person – it depends, of course – has already completely forgotten about that situation. So, hanging onto blame and situations from the past doesn't do any harm to the other person; it only harms you.

There is a famous saying: *Not forgiving, or hanging onto accusations, is like drinking poison yourself and then hoping that the other person dies.* I looked for the creator of this quote, but it seems that it could be any one of these people: Carrie Fisher, Nelson Mandela, Malachy McCourt, Emmet Fox, Bert Ghezzi, Susan Cheever, Alan Brandt, or Anonymous?

One example of this situation is Dr. Edith Eger, who survived the Holocaust in Auschwitz. She witnessed really horrible conditions and lost her whole family in the death camp. Nevertheless, she has learned to forgive these dreadful experiences and to choose love instead of hate. She consciously decided to leave her role as a victim and to forgive.

It was the same with Dr. Nelson Mandela. In 1990, when he walked out of the prison gates hand-in-hand with his wife Winnie after 27 years of imprisonment, he said, "I greet you all in the name of peace, democracy, and freedom for all." He also stated that by stepping out of prison, he was forgiving everything, because if he didn't, he would stay in prison for the rest of his life.

Do you now see the power of forgiveness? Did you ever look at forgiving from this perspective? It is quite an interesting angle, isn't it?

"Before the sun sets, forgive" is a powerful Hawaiian proverb. To forgive does not mean that you approve of what happened. It is all about letting go of what happened and what cannot be changed any more. If you have been a victim of mental or physical abuse, or have been wronged in some other way, then you can reject that act but still forgive and acknowledge the experience. You may not agree with the experience, especially if it has been traumatic, but I'm hoping that by the end of this chapter you will see the power of releasing yourself so that you can be free again.

Many people think that by forgiving somebody you are acknowledging that this person was right or did the right thing, even though it was completely wrong in your eyes. So, they choose not to forgive. But this is a misunderstanding and could be why you find it difficult to forgive.

Forgiving is not about the other person; on the contrary, it is all about breaking you free from what happened. Forgiveness means first and foremost to forgive so that YOU are free again! As long as you hold onto this blame, you continue to pay the highest price: not being able to be completely happy. See it from the perspective that you decide to love yourself more than you hate the other person.

Is forgiving something that comes naturally to you? Do you find it easy to forgive someone else?

If you forgive, it doesn't mean that you agree with the actions, but you are releasing the stored energy from your system. That whatever happened in the past doesn't have any influence on your present and future life.

You might recognize this inner dialogue, when the voice in your head asks: Do I really want to forgive that person?

No, I can't. It was not right, what he did to me. It hurt me. It destroyed my life. He doesn't deserve to be forgiven. Actually, there are hundreds of reasons why he shouldn't be forgiven, why I cannot forgive him. Why on earth should I forgive this behavior?

BUT – maybe there is also another voice in your head asking: Do I really want to carry on with this for the rest of my life? Do I want somebody else to have the power to block my development? Do I want this blocking my development because I just can't let it go?

Hopefully, these questions can help you to understand that this is about you and not about the other person. You remain trapped in your own past. If you don't let go, you are adding and adding to your emotional backpack until the moment you can't carry any more, which is when the body says enough.

So, it's better that you let go now. It's time to get rid of your old ballast and take your life back into your own hands again. To free yourself.

You deserve it!

Most of the time we are not even aware that we are carrying around blame or rage, or that we feel like a victim of what we experienced in our past. But it can indeed be life-changing to forgive, as I experienced during one session with Mary. I had met her on different occasions some years before and got to know her as a strong, confident, and very analytical woman. When she asked to have a session together, she told me that she had lost her job two years ago and had two new jobs since then, but both had ended within the first three months.

She was devastated, insecure, and had the feeling that everything she was doing was wrong. She experienced panic attacks during her jobs, and the fear of making a mistake was omnipresent. I could hardly recognize that strong woman that she used to be.

During the session I connected with her energy, and we soon found the reason behind the situation she was in. Due to her parents separating when she was four years old, Mary's mom had been forced to leave her with her grandmother in order to continue with her job and to secure the financial stability of the family. For that reason, Mary felt unwanted and left behind by her mom, and even though her grandmother did everything to make her feel at home and safe, it was not the same. She wanted her mom to be around and spend time with her, and to show that she cared for and loved her.

So, there were these huge accusations that had never been spoken about between Mary and her mother. As we discussed this and looked into it, we also tried to understand how Mary's mother had felt at that time. She felt guilty for leaving Mary behind and not being there for her, and we could both feel her guilt, blame, and sorrow.

Mary had been so occupied with her own pain and sadness that she'd never thought about how her mother felt about the situation. But by connecting with the energy of her mom, all of a sudden Mary could feel her mom's emotions, and tears started to run down her cheeks. This realization helped her to let go of all the blame and accusations she had against her mom for not being there for her when she needed her. We followed this up by using the Energy Exercise described below so that Mary could forgive her mother. It was such a touching moment to see all the unspoken words, emotions, and tears flowing out of her, and to bring harmony back into their relationship by forgiving.

At the end of the session Mary looked completely different. There was an expression of softness and peace in her face that I shall never forget. She told me immediately that she felt so much lighter in her heart and in general, as if a huge stone had been lifted from her shoulders. After a week she wrote to me: "The session with you did so much good, I am still excited. Immediately afterwards I noticed how the energy in me started to flow again. I am now much less concerned with the question of what others think of me, and I have regained confidence in myself and the world." And guess what – yes, she found the ideal job where her know-how and expertise was appreciated. Isn't it fabulous how just spending 30 minutes in your past and letting go of what happened at that time can have such a life-changing impact on your current life?

So please, try it. Let go of all this old stuff that is blocking your energy and holding you back; it is definitely not worth hanging onto. The following Energy Exercise will show you the power of forgiveness.

ENERGY EXERCISE #23 – Forgiveness always takes place in the heart and not in the mind

Even if you feel that you would like to forgive, it is still hard to walk up to somebody and tell the person: "What you did was not right, but I forgive you."

And that is exactly the reason why I developed this very powerful Energy Exercise. The charm of this exercise is that you connect with the energy of the other person (you already now know that this is possible) and forgive them energetically. This is just as powerful as talking to the person directly, and has the same result when you do it open-heartedly. Want to try it out?

Think of a situation in which another person said something or did something that hurt you, and you still haven't forgiven that person. Connect first with your heart, and then with that person energetically, then forgive. Follow the steps as they are described below:

01 Stand up and close your eyes.

02 Connect with your heart by putting your attention there.

03 Connect with the person you want to forgive.

04 Imagine that the person is standing right in front of you.

05 Imagine that a white-golden light gets from your heart to the heart of the other person.

06 Decide that you don't want to carry on that blame and accusation into your future life.

07 Decide that you want to let go and be free again.

08 Connect with your heart and forgive that person.

09 Check with the Standing Method if you have forgiven that person.

How do you feel now? Do you feel differently after you have forgiven that person? For me, forgiving is such a liberating process, and I feel so much more balanced and in harmony afterwards.

If you are having difficulty forgiving, ask yourself why you are holding onto the situation. What is in it for you? Why won't you let it go? I have to admit that at the beginning it was more difficult for me to forgive. It just didn't feel right. But then I asked myself if I really wanted to carry this situation on for the rest of my life, and I always knew it was the right decision to let it go.

You can use the same methodology also to forgive yourself, or a group of people. And you don't need to do this exercise one at a time over different days. You can repeat it again and again as often as you want. You can block out some time in your diary, if you want, and do a Forgiveness Day.

It is very powerful and liberating to forgive, and you can never do it too often.

Ho'oponopono

Ho'oponopono is a traditional Hawaiian practice of reconciliation and forgiveness, and it means "to put in order". Through forgiveness, Ho'oponopono offers the possibility to find the way of your heart and thus to walk the path into peace and trust. When you notice that old resentment or anger comes up in your everyday life, repeat again and again the following four phrases from the Hawaiian forgiveness ritual Ho'oponopono:

01 **I am sorry.**

02 **Please forgive me.**

03 **I love you.**

04 **Thank you.**

01) I am sorry: With this sentence we acknowledge that you did something wrong, or something did not go as it should. The moment we speak this sentence, we acknowledge a mistake and no longer deny it. We accept the situation as it is and acknowledge that it has caused us and others suffering.

02) Please forgive me: With these words we ask the other person, and ourselves, to forgive us for the pain we have caused.

03) I love you: I love you and I love myself with all my strengths but also weaknesses. Through these words, we acknowledge ourselves and others, and restore harmony.

04) Thank you: We say thank you that the power of forgiveness heals, and that we are allowed to let go and heal.

Ho'oponopono is a powerful forgiveness ritual. It helps you to restore order and harmony in yourself and your environment, especially if you did something wrong and you want to ask the other person to forgive you.

I had heard about Ho'oponopono a couple of times, but the first time I really used it was mind-changing. It all started with an emotional discussion with my older daughter. I wanted her to do something, but she had already decided that she didn't want to do it and ran off to hide in the bathroom. I didn't like the way she handled the discussion, and just running away was certainly not the way I wanted her to unwind from this situation. So, I ran after her and put my foot in the door to prevent her from locking herself in the bathroom. Exactly at the same time, she slammed the bathroom door, and before I could react my foot got literally squashed by the door. I felt the immense pain when the door hit my foot, and my scream of pain made my daughter jerk open the door again.

I immediately put some ice packs on my foot to keep it cool and started some energy healing work. Once I felt a little better, I asked my daughter to sit down with me at the kitchen table so we could forgive ourselves and let the whole situation go. It was important for me to sort out this situation at that time. I didn't want to hang onto all the anger, pain, and blame. The same was true for my daughter. She felt scared and guilty because she had hurt me – even if it had not been her intention.

So, we sat down at the kitchen table, held hands, looked into each other's eyes, and said the four sentences – one after the other – to each other. It was a very special and emotional moment, and we both felt so connected and tearful. After the last sentence, we hugged each other, and we both felt so relieved. It is hard to describe, but we both felt that things were okay now for both of us, that everything had been cleared out energetically. That was the moment when I realized how powerful this little ritual is. It immediately changed the whole situation and the energy between us.

The most amazing part of the story is still to come. Later, when I finally lay in bed, I noticed that my foot already hurt much less, which I was very grateful for. And when I woke up the next morning, I checked my foot, and it was almost back to normal. To my surprise, there were no bruises and almost no swelling, so I could even wear my normal shoes.

This incident taught me the power of forgiving! And this ritual of Ho'oponopono has changed my life completely. That's why I show you another powerful Energy Exercise which you see right below, with which you can integrate forgiveness in your daily life. It works very simply, and I now use it daily.

ENERGY EXERCISE #24 – Ask for forgiveness, forgive others and yourself

There is another very powerful Energy Exercise[6] which I use every evening before I go to bed. I stand right next to my bed, close my eyes, and mentally do the following ritual:

01 **I ask forgiveness from all the beings that I have hurt during the day by what I have said, done or thought.**

02 **I forgive all beings that hurt me during the day with what they said, did or thought.**

03 **I forgive myself for everything that I said, did or thought during the day that hurt myself.**

It is a very powerful ritual to clear the energy before you go to bed and to close the day. Sometimes we do, or say, or even think things unconsciously that

---

6   I learnt it from my mentor Stefan Klitzsch.

might hurt other people, or we are hurt by what others do or say to us.

This ritual supports your willingness to have an easy-going and forgiving attitude towards life. And to let everything go before you go to bed helps you to enjoy a more refreshing and restful sleep. You will feel the difference!

Your Energy Insights

Most of the time there is one thing that keeps us from forgiving others: when we think that what the other person said or did was not right. And that's true; it probably wasn't. But be honest with yourself. Do you really want to carry this around with you for the rest of your life? Do you really want to give somebody else the power to keep you angry, sad, or hurt? And even worse, it's likely that the other person has already long forgotten about that situation, yet it is still haunting your life.

I know it sounds difficult, and obviously I don't know exactly what happened to you. But once you have experienced the power of forgiveness and how liberated and full of energy you feel afterwards, you will ask yourself why you didn't do it much earlier. You might even blame yourself for not forgiving and letting it go sooner.

Forgiving is all about transforming this stored energy that is blocking you, and shifting the perspective of how you look at a certain situation or person.

The situation is still there in your past – but it won't influence your future anymore.

The most powerful aspect of forgiveness is to clean yourself and your energy body from energies that do not belong to you. Believe me, once you get started, you will find it so liberating that you won't be able to stop. And when you change YOUR energy, it will have an impact on people around you. Just be open and observe how people act differently towards you, just because YOU have changed YOUR energy.

Chapter Eleven

Let go

LET GO AND
BREAK FREE

"Be the energy you want to attract."
Korra the explorer

Let go

Have you felt frustrated when you want to move on but can't, and something keeps coming back?

Do you feel frustrated when you have tried to move on from a situation that has triggered you, and it keeps coming back to repeat itself?

I experienced it myself. Every once in a while, something popped up that I thought I had moved on from, and this frustrated me a lot. I had already looked at it and worked on it, but it kept coming back! What was wrong with this? And even worse: What was wrong with me? It was so frustrating to want to move on yet be unable to figure out why I was still feeling stuck.

It took me quite some time to work out that there are different reasons why things keep coming back to you. One might be that you haven't solved or healed the real root behind it.

Just a couple of months ago, a friend called and told me that she had been let go by her company. We talked about the situation and how she felt about it, and I tried to comfort her. I fully understood how she was feeling, because a couple of years ago I had been in the exact same situation. The company I was working for at that time went through a difficult period after a merger, and people were forced to leave the company. As I had expected the worst, in a way I had prepared myself for this happening. But when it actually happens to you, everything changes.

I still remember the exact moment when I was called into that room. It was such a weird situation when I got the news, and I was overwhelmed by an avalanche of conflicting emotions. On one side I felt relieved that it was finally over, and I could escape this space of fear and negative energy. On the other side I felt hurt and also ashamed at being made redundant and not being needed any more.

The moment I hung up the call with my friend, all of a sudden – and almost out of the blue – I started crying, and I felt immediately dragged back into what had happened to me at that time. I realized that there were still some emotions that were stored in my system that wanted to be seen and let go. I could feel all the sadness and shame that I hadn't allowed myself to experience at that time. Back then, these emotions had drained me and made me feel so bad that I had just pushed them away.

Now, looking back at my career and knowing that I secured a much more impactful and powerful position afterwards, I know that I don't need to have these emotions anymore. On the contrary, I now know that I don't need to feel ashamed and sad at all, because this change turned out to be so much better for me. In all honesty, if I had stayed at the old company much longer, it wouldn't have been good for me in the long run at all.

So, I asked all these stored emotions to show themselves. I felt them again for a split second, and then I let them go. I stayed in the situation and asked myself if there were more things that wanted to go, and I felt a sharp pain in my heart. I healed it and let it go as well. Then I checked and realized that there were also some beliefs that were holding me back, which were not suitable for my life anymore. So, I cleared them out as well. Finally, I accepted and took on the responsibility that this situation had happened to me.

It was such a relief. I had not been aware that I was still carrying all these emotions and the burden of the whole situation with me. Afterwards, I felt free and so much lighter, and my future looked brighter. It is always the same. After letting go, I actually feel stronger and more powerful. I want to share my secret Energy Exercise so that you can do the same. And in this context I also learned another important lesson: Did you know you can only let go of something when you accept and take on responsibility for it? More about this in Chapter Fourteen.

The Wheel of Energy Blockades

In order to let go, you need to know what you need to let go of in the first place. Only afterwards can you move on. In order to make this process as simple as possible for you, I have developed the Wheel of Energy Blockades. When working on a certain topic, situation, emotion, or trigger point, you can use this wheel to find out the actual reasons and causes keeping your energy from flowing optimally. By checking with the Standing Method, you can find the energy blockades then eliminate them afterwards.

Before I introduce this powerful process to you, I shall give you a short explanation of each section of the Wheel of Energy Blockades.

As a kind of overview, have a look at the Wheel below:

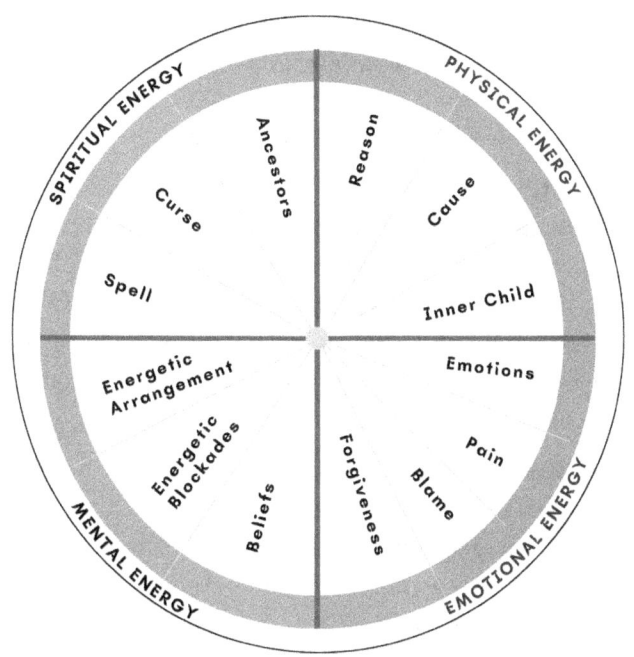

Quarter 1 – The Physical Energy

As you can see, there are four quarters in the Wheel. The first quarter is related to finding the reason and cause behind your problem. Many situations or conversations that are triggering you are not triggering *you* but your Inner Child. Resolving issues with your Inner Child is a special process which I will explain to you later in the chapter.

Quarter 2 – The Emotional Energy

The second quarter deals with detecting stored emotions as pain, fear, and blame, and how you release them and forgive yourself or other people. This is such a liberating process. It is so incredibly simple, and you feel like a new person afterwards.

Quarter 3 – The Mental Energy

The third quarter is about letting go of limiting beliefs (we already spoke about your beliefs in Chapter Nine – do you remember?). If you haven't started replacing your limiting beliefs with empowering ones, NOW is the time to do so. Here you are dealing with those limiting beliefs within the context of the situation that you are working on.

Other common blockers are energetic arrangements and blockades. What is hidden behind these terms? Energetic arrangements are promises, (wedding) vows, oaths, or permissions you have given other people at a certain time in your life. This may have happened when you were a child, but they can still be valid even though you are not aware of them anymore. Nevertheless, they can block your energy, and that's why it is so important to clear them out. Imagine that you feel energetically still bound by your wedding vow to your ex-husband, but you are now married to somebody else. Do you really want that?

Quarter 4 – The Spiritual Energy

The fourth quarter brings your attention to the fact that, unfortunately, human beings are not always kind to each other. I am quite sure that you can relate to situations in which you were annoyed by other people and thought that you would like to shoot them to the moon, or any other not-so-nice thought. These negative thoughts from other people – depending on the intensity – can settle in your energy body as curses or spells and block your energy from flowing. And you don't want that! That's why it is good to check if you need to release them from your energy body as well.

Just by thinking about it and being more aware of it, try to stay kind with your own thoughts and feelings towards another person, even though this person is annoying you. It works vice-versa: You don't want bad thoughts in your system, so don't send them to other people as well. A curse or spell can be a real energy blocker, and not only in Harry Potter!

And last but not least – yes, your ancestors can also block your energy and development. Many beliefs, thoughts, and emotions have been passed onto you by your genes without you being aware of them. If you don't believe me, let me share a fascinating research study in this context.

A couple of years ago I heard about an experiment with rats[7]. Let me make one thing clear here, I am completely against any animal experiments. However, the result that scientists found was so surprising that I would like to share this story with you. During the experiment, rats were taught to be afraid of the scent of cherry blossoms. They were trained to associate the scent of cherry blossoms with the fear of receiving an electric shock. The fascinating part of this story is that the next generation of rats were also afraid when

---

7 This study by Brian Dias at Emory University School of Medicine in Atlanta in 2013, provides evidence for the inheritance of memories or traits across generations.

they smelled that scent, even though no electric shock was imposed on their generation. So, their parents had passed on this fear to their children. And even more shocking: the same was true with the next generation. The rats appear to have inherited the fear knowledge through modifications to their genetic code.

What does this mean for you? Have you ever found yourself behaving in a certain situation in the way that your parents or even grandparents did? Have you ever wondered how you came to adopt these behavioral patterns? The answer may lie in the inheritance of their emotions or beliefs.

As research shows, when we store emotional states like stress, anxiety, or depression in our bodies, we can also pass them down through multiple generations. When I became aware of the power of my ancestors, I had the very strong intention to interrupt this process – not only for myself, but most importantly for my daughters. If this is also important for you, you must try the Energy Exercise I am showing you in this chapter.

But before we dive deeper into this Energy Exercise, I promised to explain a little bit more about the different parts of the Wheel of Energy Blockades – like, for example, your Inner Child.

Triggered and you don't know why?

Are you sometimes triggered by a certain situation or specific circumstances, or just because somebody said something to you? Somebody pushes a certain button and then a kind of automatic program starts to run and unwinds, and there is nothing in the world that can stop it. I guess we all know such situations.

I remember the trigger story from my client Mary. The first time we met she was completely irritated by the school system. She complained a lot about the school system itself, the teachers, how her kids were treated, and that they weren't learning the right things. My daughters were going to similar schools, but although I am not a big fan of that system either, I wouldn't care to grumble about it for hours. I accept that the school system is like this here, and I am grateful that my daughters have the possibility to go to school, even if they don't learn things they really need for life.

I pointed that out to Mary, and I think that was the first time she reflected on it and understood that most people around her were not triggered by the school and teachers in the same way as she was. That's when she realized that her behavior was costing her a lot of energy and didn't lead to any change.

The reason for these trigger points and your irrational behavior is that this is not the *adult you* reacting; it is your hurt Inner Child. And this will continue until you heal the reason behind it. When you were a child, probably some situation happened which made you angry, felt ignored, not seen, ashamed, overruled, and much more. As you were overwhelmed by certain situations, emotions, and people around you, you just stored them in your energy system. And whenever somebody triggers a similar situation nowadays, you go back into that moment of how you felt and reacted as a child. So, most of the time when adults are quarreling with each other, it's actually the little Inner Children quarreling with each other. Very scary thought, isn't it?

And even worse – most people around us (especially our family members and friends) know us so well that they (unconsciously) know our trigger points. Not only that. Our parents, and to a degree also our siblings, even *programmed* our nervous system and trigger points. This is like giving the remote control to your trigger points to somebody else and allowing them to push the buttons and select the program. Do you really want that?

The only way out of this is to heal these trigger points and automatic programs. And I can promise you, it is so liberating and such a wonderful feeling when somebody presses one of your trigger points and suddenly you react in a completely different way… or not at all.

ENERGY EXERCISE #25 – Talk to your Inner Child

It is so important to get in contact with your Inner Child and release these trigger situations, because in many situations, as a grown-up adult you would react in a different way altogether.

So, when you find a situation in which your Inner Child was hurt, follow this process:

01   Stand up and close your eyes.

02   Connect with your Inner Child by visualizing yourself as a child.

03   Imagine that you take it in your arm or put it on your lap.

04   Talk to your Inner Child and ask it questions such as:
- How are you feeling right now?
- What exactly caused this feeling?
- What do you need right now?
- Is there anything I can do to make you feel loved and happy?

05   Just be with your Inner Child, perceive it, feel its hurts, fears, anger or even shame with your full attention and free of judgments.

06   Allow the feelings to come up. If you want to cry, allow yourself and your Inner Child to cry.

07   Give your Inner Child everything it needs right now.

08   Consciously decide to take the responsibility to heal the hurts of your Inner Child and to reawaken the love and trust in it.

09   You can tell your Inner Child that you are always there for it when it needs you and that you will take care of it from now on.

10   Ask your Inner Child if the situation is healed now and if there is anything else it needs.

11   Check with the Standing Method if the situation is resolved by 100%.

If you haven't been in contact with your Inner Child, there might be the possibility that it is hiding (behind the sofa) and refuses to come out. If this is the case, please stay in contact and persuade it to come out and start communicating with you. Try to create a safe space. When I finally got mine to say anything, the first and only thing she said that day was, "I want YOU to be my mommy." Your Inner Child longs for your attention and love, so allow your Inner Child to recognize itself in its wonderful uniqueness and to feel loved and protected.

If you have the feeling in your everyday life that you are very emotional, angry or anxious, without knowing exactly why, this is a really good opportunity to connect with your Inner Child. Ideally, you should even spend time daily with your Inner Child.

Imagine meeting with your Inner Child in one of your favorite places. Then start a conversation with him or her by asking questions like: Did this moment remind you of an experience from our childhood or trigger something in you? Does this person right now remind you of someone in your immediate circle (mom, dad, siblings, grandparents)? Would you like me to do something different in my life right now?

You can support your Inner Child by providing it with reassurance, like: It's perfectly okay to make mistakes. I love you no matter what you do. Or, I want you to know that I am here for you, I love you, and I will never leave.

Are you brave enough?

Are you brave enough to let your past go now? Are you courageous enough to leave all causes, reasons, and blame behind, and give yourself the freedom that your present and future life do not depend anymore on your past? Scary thought? Yes, but it is also liberating, isn't it?

Before we dive into how YOU can do it, let me tell you about my client Randia. She came with the request to me that her energy needed more structure. She explained that she always liked to have extremely well-organized places, both at work and at home, because she needs structure in her life. But at that time her whole working space, she told me, was a complete mess with papers all over the place, and she just could not manage it anymore. She didn't even show me, because she was too ashamed about how messy and chaotic her office space looked. In her mind she just needed a little more time to get it organized.

But of course, there was more behind it than just having enough time to clean up the chaos. So, we looked into her mess, or to put it better, into what habits were keeping her from organizing it. We tested it out with the Standing Method and got answers that there was one cause from this life and a reason from a former life.

We decided to begin with the cause from her current life and continued testing along the Wheel of Energy Blockades, then removed a huge energy blockade. We both felt the incredible energy from that blockade, because all of a sudden, we felt very hot. The second step was to release some pain from her past.

Randia wanted to know about what had happened, and we finally found out that we were working on a situation that occurred when she was about five years old. Somebody had complained that she wasn't putting her toys away, but it was not her mother or anyone from her family, so it might have been when she was visiting a friend. The young Randia had been hurt by the tone used when this person spoke to her, and she had stored that pain in her system. But now she had the opportunity to decide that she didn't want to hold onto this pain anymore. So, she connected with the pain, allowed it to show itself for a short moment, then healed it and let it go.

The third thing we looked into was an energetic arrangement. When she had been about the same age, she had promised something to another child who was her best friend at that time. But both were now adults and in completely different situations, so Randia was at liberty to terminate this agreement. She realized that what had been appropriate when they were young children was not the same anymore, so it made no sense for her to be energetically bound to that situation. We opened the space to let go of this agreement and terminated it, because it was no longer suitable.

As you can see, we didn't look into too many details, because most of the time it is not really necessary. The reason for this is that usually the situations that show up are completely banal when looked at from today's point of view. As a child you may well have felt hurt in certain situations, or overwhelmed, even though most of the time these situations were rather trivial. However, at that time you perceived the situation in a different way, and that's why you held onto these emotions and stored them in your system. For this reason, most of the time you don't need to know exactly what happened because it was not really important. But there might be important information you need to look into more closely to let it go and learn from the experience. That's why it is always handy to check with the Standing Method if you need to know more. Most of the time you won't.

You may be shocked that now, after a couple of scientific disclaimers, I am now adding a spiritual disclaimer, but it is possible for scientists to also be spiritual. Who would have thought that? However, the logical part of my brain was wondering whether to include this next part of Randia's journey in the book, as perhaps it was too far from our "normal" thinking. Many scientists do not recognize that there is so much more that we cannot grasp with our brain, but that doesn't mean that it doesn't exist. So, the battle between me being a scientist and my spiritual part went on for a while... and was won by the spiritual part of me.

The second reason for the chaos in Randia's life was due to something that had happened in one of her former lives. In that life, Randia had been a woman, not yet married, when she overheard something that somebody said about her, and which really hurt her at that time. It was something rather trivial, which is why we wouldn't have needed more information about it. But Randia was curious, so we found out that a woman had said something negative about the former Randia, mocking her dress and saying how ugly she looked in it.

Naturally, Randia didn't want to carry on holding onto this situation, so she decided to connect with her heart and forgive that woman for whatever she had said about her at that time. It was not important anymore, and that's why it was easy for Randia to let it go.

Afterwards, there was nothing more to do, as it seemed that the situations had all been cleared out. I asked Randia to observe how things developed and to see if she suddenly got the urge to put everything in its place and bring more structure in her life.

At the end of the session, Randia felt relieved. She could feel that the blockage had been removed, and she was ready to deal with all her papers and write important things down. She saw that the chaos was still there, but now felt ready to deal with it.

Two weeks later, when we had another call, I was amazed when Randia showed me around her room. It was completely tidy, all the papers were gone, and the whole room looked comfortable and inviting. It was unbelievable, and I could feel how proud she felt and how relieved she was that this nightmare with her chaotic office had been solved. She told me that her daughter had helped her to sort all her papers, throwing some of them away and filing the other ones in different folders and drawers. There was such a positive energy in the room. On top of that, Randia showed me on the screen a mind map she had created about how she wanted to organize her life in a better way. I was really impressed. Once again, it proved how powerful Personal Energy Management actually is.

Another interesting part of Randia's story was that the day after our session (she told me afterwards), her best friend from childhood called her. They had lost touch because Randia had moved away, but all of a sudden (there are no coincidences, right?) the woman called. And at the end of the phone call, she invited Randia and her daughter to visit her in her home in France, because they hadn't seen each other for 14 years. Isn't this incredible?

ENERGY EXERCISE #26 – Clear your energy blocks and let go

As already shown through different examples, with this process you can clear out energy blockades from this or former lives, or in your ancestor lines. By testing with the Standing Method, you can find out the cause or reason of a certain behavior, or release a trigger point. In the course of the process, you can let go of stored emotions and pain, let go of blame, or forgive yourself or others. It is a very powerful process.

The Energy Exercise is based on the Wheel of Energy Blocks which I introduced to you by explaining the different quarters and sections in detail. So, when you start with the Energy Exercise, you might find it helpful to look back at the graph showing the Wheel of Energy Blockades at the beginning of the chapter.

Now it is your turn! Just go for it! You will be rewarded by feeling so much lighter afterwards. Don't you think this alone is reason enough to try it out with the Standing Method? Just start with a problem or situation that is really bothering you and find out what is the cause or reason behind it. Be open to how things evolve, and follow your energy. It will guide you in the right direction.

01 Stand up and close your eyes.

02 Ask with the Standing Method if it is an Internal or External Problem, or if it is coming from your Ancestor Lines.

03 If it is an External Problem, heal it, bring it into the solution or let it go.

04 Check with the Standing Method if the External Problem is solved.

05 If it is an Internal Problem then continue with the following steps:

06 Find the reason behind the situation or trigger, by asking if it is a cause or reason
- Reasons are from former lives.
- Causes are from this life.
- Check out if it is one or several reasons or causes.
- Then heal them, bring them into the solution or let them go.

07 When it is a cause, check out if you need to heal your Inner Child as well:
- If YES, connect with your Inner Child.
- Ask how old you were when the situation happened.
- Remember yourself what this might be.
- Ask what your Inner Child needs.
- Give your Inner Child what it needs and wants.
- Ask if it wants to let go of emotions or pain.
- Let go of any stored emotions or pain together.
- Ask if you need to let go of blame and forgive somebody.
- If YES, please do so.
- Check if the situation of your Inner Child has been fully resolved.

08 Have a look if there are some emotions or pains stored that want to be set free
- Check if there are emotions or pains stored.
- Ask them to show themselves to 100%.
- Decide to let them go.
- Sometimes it is necessary that you need to let go of blame or to forgive someone or yourself before the emotions and pain can go.

09 Ask if you need to forgive somebody
- Yourself.
- All others.
- Your ancestors.
- Life.
- The Universe.
- God.
- Then connect with your heart and forgive.

10 Check if there is a blame connected with your need to forgive somebody. Decide to let go of any blame and that it doesn't have any impact on your life anymore.

11 Have a look at your beliefs
- Are there any beliefs that don't serve you anymore?
- Let them go, heal them or cancel them.
- Ask if you need to replace these beliefs with new ones.
- Think of three new beliefs you want to implement and activate them.

12 Find out if a spell, curse, energetic blockade or energetic arrangement is connected to it.
- When you find it is one of those, ask if it is one or several of these things.
- Then end it, cancel it, declare it invalid or not applicable anymore, resolve it.

13 If it is a problem related to your Ancestor Line, check any issues with your Ancestor Line
- Ask if you should look into your mother's or father's line.
- Ask if you need to go back 1, 2, 3, 4, ... generations.
- Ask if it is woman or man.
- Ask if you need to know what the problem is.
- Ask if you need to forgive your ancestors.
- Ask if there are any emotions or pain stored from your ancestors.
- If NO, ask if you can heal it, solve it or let go.
- If YES, ask your ancestor how you can solve the issue.
- Send white-golden light down your ancestor line.

14 Check with the Standing Method if the situation in your Ancestor Line has been dissolved.

15 Finally, check if you have resolved the whole situation to 100%.

16 Fill the space you liberated, because you cleared out so many things, with divine energy.

Little note: Sometimes you need to unlock one thing before you can let go of something else, because they are connected. And often there is an order in the process. What do I mean? For example, sometimes you need to forgive before you can let go of a pain. Or eliminate a spell before you can let go of emotions. So, if the letting go process doesn't work for you, check whether you need to do something else first.

Bear in mind, too, that at times you might need to do several rounds, because the information is stored in different layers – like with an onion. When you let go of pain and emotions, all of a sudden another reason, cause, or spell evolves, and you can work on it.

Therefore, when you think you are done, it's best to make a short check on the four quarters of the Wheel of Energy Blockades to make sure you worked on everything that you can right now, and make sure everything has dissolved.

Please don't be too hard on yourself if you get stuck or you cannot move forward with the process. Simply ask your mind to step aside for a moment, then connect with your energy. Trust that it will show you the way. Just follow your intuition, and if something feels right for you just do it, despite what is written here in this Energy Exercise. This Energy Exercise is just a process to get you started, but you are special and so is your energy. So, there will be situations where you will do something differently and it will work. Trust yourself!

Sometimes you might have to change the order of the different steps, but this does not mean the process doesn't work for you. We are all unique and have different frequencies modulated on our frequency, so the situation may well be different on your side. But this is all part of your journey!

For me, it was amazing to watch my client Ria do it the first time. I guided her through the process and made suggestions about what she could ask and how she could react to what came up. Then after a couple of minutes, she took over and just followed her energy about what to look at next and how to deal with it. It was so powerful to watch how she took on the ownership of getting her energy back. I just held the space for her, and she finished the whole process smiling and so proud of herself at how easily she had managed to remove the blockades of the situation she was working on. She is the expert in healing herself, though, and so are YOU! If she can do it, you can too! So, connect with your energy and get started!

Don't worry if the same topic pops up again at another time. You didn't do anything wrong, and it is quite a normal experience. Depending on what it is and how much it hurt you, there can sometimes be different layers to it, and you cannot work on all of them at once. But nothing will show up that will be too big for you to handle. If it is big, it will appear in little tranches (like a salami tactic), giving you enough space to maneuver, to manage, and to let go. This very powerful self-protection mechanism, which we all have, gives you the assurance that you can handle everything that shows up.

However, there are times when you might get the impression that you are continuously dealing with the same topic, and it keeps showing up again. I feel you! For a long time, one topic in particular in my life came up over and over again. I really got mad about it and thought I was stuck with it for the rest of my life. In truth, I still haven't fully solved it, but now I can see that I have moved on and things have changed. So even if you sometimes get the impression that you are stuck – you aren't. You just haven't been able to heal the real cause or reason or root behind the issue, but you will get there eventually. I genuinely believe that, which is why I keep going with my topic. So please, I beg you to keep going! It is definitely worth the hassle.

Life works in a spiral. You never meet the same issue twice. Just a different layer on a different level...

How energy management works at its best right when you need it

Just to give you another example of how energy management works, a couple of weeks ago I was on vacation with my family. We went by train and had a stopover in Milan. I went to get something to eat for my daughters but ended up being completely upset.

There was a long queue to buy some small pizza pieces, so I stood and waited, but when it was finally my turn I realized that all the vegetarian options were sold out. I was disappointed, but then I saw one of the ladies bring out a new baking tray of Margarita pizza. My face lit up and I asked her if I could have two pieces of those. I got a clear no. These pieces, I was told, were made to be sold with meat on top and were double the price of the vegetarian version. So, either I would have to pay the higher price, as it would be with the meat topping, or I couldn't buy them. My daughters and I were really hungry so I had no other option, but I was so angry I could hardly speak. If it had just been for me, I wouldn't have given in and bought the pizza pieces at all.

When I got back to my daughters and handed over the pizza, I realized that I was still in a rage. So, I started looking into my anger and where it was coming from. Of course it was ok to be angry, but my anger was huge and not normal, and so it was best to look into where it was coming from and try to solve the issue immediately.

First, I asked with the Standing Method whether the cause behind it was coming from my own childhood. Immediately I saw myself as a little girl at one of the various events where I had to join my parents. They had a lot of friends, and every weekend there was either a party at our place or they were invited to a party at one of their friend's homes. Most of the time they took me along and I would spend hours being invisible, reading one of my many books until I finally fell asleep, then they picked me up when they finally went home. At that time, I was never asked if I liked being there. Of course, sometimes there were other children around that I could play with, but I always felt that what I wanted never really mattered.

Now I understood that this anger was not coming from the actual situation at the pizza shop, but from the seven-year-old me who was angry that once again her wishes and needs had not been taken into account. So, I connected with my seven-year-old Cornelia, and she explained to me that she didn't want to feel angry but just wanted her needs and wishes to be seen and to be shown appreciation. We spoke about the situation in the past and what she would need to feel more comfortable. Then I pulled her onto my lap, hugged her, and gave her all the love she was longing for. I gave her everything she needed to be felt seen and visible, and together we decided to let go and allow all anger and other emotions to flow out of our system. We didn't need them anymore.

Then we had a look at whether there were some limiting beliefs we should clear out, which we did, and finally we connected our hearts with the hearts of my parents and forgave them. Altogether, this process didn't take much longer than ten minutes, and afterwards I felt completely calm, balanced, and full of energy again. All the anger was gone, and I was so glad that I had taken the time to look into it immediately and to shift the energy. It is always amazing how simply this works. You just need to look into the issue and let go, and you will feel so much lighter afterwards.

This is just an example of 1) Emotional First Aid; and 2) That every now and then something can pop up when there is a new trigger situation. It just keeps going, and here I show you how you can deal with it on the spot.

Your Energy Insights

You may feel that the goal of mindset work is about becoming a different person, which I mentioned at the end of Chapter Nine. But actually, managing your energy and letting go is all about *becoming the person again that you actually are*. Because the person you *are* right now is not you! It is a product of all your emotional and mental baggage that you have been carrying around for such a long time that you have forgotten how it feels to be the REAL you. Energy management is all about letting go of what is NOT you and rediscovering your REAL you.

At the beginning, the Wheel of Energy Blockades and how to use it may look a little bit overwhelming and complex. But you want to dissolve your energy blockades, don't you? Looking into any situation that bothers or triggers you, then clearing it, results in one frequency less on top of your unique frequency! And if I can do that, you can too! Own your power! Own the power of your mind and energy!

Letting go is all about letting go of the past, and leaving it where it belongs – in the past.

Chapter Twelve

Connect with your heart energy

CONNECT WITH YOUR HEART ENERGY

"Love is more than a feeling of affection and attraction - It is an energy source."
Debra L. Reble

Connect with your heart energy

Now we are going to look at your heart's energy. Are you intrigued, scared, excited? Or maybe all together?

Do you ever wonder what your heart is for? Sure, it is an important organ, a pump whose sole purpose is to circulate blood throughout your body, and it works hard beating day in and out. But besides that, what's so great about it? Or do you have a romantic association with your heart? Do you expect it to always be soft, loving, and understanding?

Your heart, though, is so much more than a pump. It's much more than the place of your warmest emotions and deepest feelings. In fact, as science is beginning to reveal, the heart is much more powerful, healing, and transformative than has ever been imagined. That's why we are going to look into your heart from an energetic perspective. And you can be sure that whatever you discover here, it will definitely change the way you look at your heart – promise!

Warning: Here is another (cheeky) science disclaimer.

As always, I am only sharing exciting stuff that you *have to* know! From the energetic point of view, your heart is the center of your body. In quantum physics, it is called the second brain. Research has proven that more impulses and information are sent from your heart to your brain than vice versa. The magnetic field of your heart is *5,000 times stronger* than the one from your brain, and the electric field 60 times stronger. Hence, your energy body is mainly built by the electro-magnetic field from your heart. That's why it is so important.

I was completely blown away when I learned that, and it actually turned my world upside down. Until then I had been so focused on my brain and thoughts, but then I realized that connecting with my heart – and its much bigger energy field – is so much more powerful. At the HeartMath Institute in the United States, this very phenomenon was studied, and it was discovered that the heart, and hence our energy body, has the ability to anticipate events in the near future. And that's why our heart and energy are so good at helping us to make the right decisions. For everything! For making decisions, talking to people, and yes, even writing this book. Everything you are doing, thinking, and feeling gets such a different quality. When you do something from your heart, you FEEL the difference, don't you?

Let's get back to science. Your heart has the strongest electro-magnetic power in your body and supplies your entire system with energy and information. On top of that, your heart energy (especially on whatever frequency it is vibrating, which you know now) determines which people and circumstances you attract into your lives and which you do not. Thus, your heart influences the extent to which you live your lives with joy and ease. So, the key is to face the world with a high heart frequency and an open heart. Have you ever hesitated to open your heart because you have been hurt in the past, or have witnessed the emotional pain of others? I am quite sure this has happened to you at some point. In our misconception of security, we often close our heart to keep it from further pain. We draw high walls around it so that we cannot be hurt anymore. Unfortunately, though, this behavior hurts exactly one person in this world – us.

Your heart is strong and powerful and longs for adventure, for new experiences, and for real closeness. An open heart is the only way to truly allow love to enter. Let's look more closely at how open your heart is.

Discover your heart frequency

You are probably wondering why on earth it is important to know your heart frequency. And you are right, there are most likely more important things to know.

It is similar to stepping on a scale each morning and knowing how much you weigh. It is not the number that is important, but what you do with this information. Let's have a look at your weight indication. What do you do with this information? Does it make you happy? Do you feel annoyed? Is it going to have an influence on what you are going to eat that day?

It is the same with knowing your heart frequency. The information creates more awareness about where you stand right now. If you want to understand what I mean, go back to Chapter Eight, in which I describe the Scale of Emotions and what impact the frequency you are currently vibrating on has on your life. Did you skip that part? Don't worry – if I was not an electrical engineer, I would have probably done the same. But you can always go back and catch up!

Coming back to your weight and to your heart frequency. It is good to know where you are standing right now in order to change it and to measure the development. It can even be fun to track the improvement, just like seeing your weight going down – or staying the same – each morning you stand on the scales.

ENERGY EXERCISE #27 – Find out your heart frequency

With this Energy Exercise you learn to track and read your heart frequency. The higher your heart frequency, the more powerful your energy body.

01 Stand up and close your eyes.

02 Connect with your heart by letting your attention flow into your heart.

03 Ask your heart on which frequency it is vibrating right now.

04 A good starting point to find your heart frequency is 100 Hertz. Ask yourself if it is more than 100 Hertz.

05 If you get a YES, ask yourself if it is more than 110 Hertz.

06 Keep going until you get a NO.

07 The number before the NO is the frequency your heart is vibrating on.

If you have a notebook, it would be helpful to write down your actual heart frequency. Just as with body weight, there is no need to test your heart frequency every day. However, every other day it would be good to have a quick check to see the change.

I guess now you are wondering how you can shift, raise, or change your heart frequency. Well, the best way to raise it is by healing your heart wounds, then forgive and let go.

ENERGY EXERCISE #28 – Feel your heart and communicate with it

Are you repeatedly ignoring the signals of your heart? Maybe you are doing this unconsciously, but they keep creeping into your life, and you are ignoring the voice of your heart. Can you really claim to feel your heart? Can you follow your heart even though your head is telling you to fear and doubt?

Feeling your heart is all about connecting to your heart and consciously perceiving its voice. So, the idea is to focus (your thoughts) on the energy of your heart, and spend a few minutes each day to connect with it by using the following Energy Exercise:

01 Sit or stand comfortably and close your eyes.

02 Bring your attention to your heart and imagine a wonderful, warm, healing light radiating from your heart into your entire body.

03 Connect with the love in your heart.

04 Ask your heart what it wants you to show today.

05 Be open to whatever emotion shows up. Whether it speaks to you in sadness or fear, delight or pleasure, its message is exactly what you need.

06 Notice the emotions and then let them flow out of your heart and out of your body.

07 When you feel your heart is lighter, free, or filled with love and joy, open your eyes.

Imagine how you look at yourself from the outside.
Look at yourself with compassion and love.
Look at yourself as you would look at a baby.
Look at yourself as the person you are.

Allow yourself that you had negative experiences and made mistakes, because that is part of life. Acknowledge yourself for all the times you have gotten back up and, above all, recognize yourself for being here right now. Connect with yourself and honor yourself for your willingness to return to yourself and your true you, and for opening your heart again.

Connect to your dreams and desires that you may have repressed throughout your life.

Do you really need these heart walls?

Throughout life, we build up heart walls through various incidents. Often, we experience emotional pain, and in such moments we build an energetic wall around our heart to prevent being hurt further. We consciously or unconsciously think that we then cannot be hurt anymore, which is unfortunately not true. A heart wall is an energetic protective layer around the heart. In principle, it is a good thing, because it protects you for a certain time while you take care of your injury, to heal.

An injury can happen when another person hurts you. But also when you hurt yourself, perhaps by acting against yourself, your values and beliefs, and also your heart. This can often happen because we have forgotten to listen to our heart, to our frequency, to our wisdom, and to our aliveness, and to act accordingly. We have forgotten to follow our heart's path, our heart's power and energy. Mainly this happens out of fear of new injuries, of the unknown, of fear of our own strength and potential. But also, out of fear of life, consequences, and responsibility (the list is endless here and different for everyone), we often leave the heart wall in place after healing, to protect our heart from being hurt again.

Heart walls only want to protect you, to make you invincible. However, your heart is sooooooo strong! You don't need to protect it.

Speaking about heart walls, I would like to share the story of Maja with you. When we met, she told me that she was in a relationship with a guy who had two children. She explained that she felt it was quite difficult for her to open up to him, and she had the feeling they were drifting apart. During our conversation she acknowledged that she was aware he was not the right partner for her, but she felt bad about ending the relationship. She just couldn't bring it over her heart.

So, we worked on what was holding her back from letting her partner go and opening herself up to a new relationship. We also looked at eliminating her belief that she was scared to accept love. She wanted to be loved as the person she was but had a huge fear of showing herself that way because she thought she was not loveable enough. Then, most importantly, we removed her three heart walls.

What happened afterwards? Right after our session she was still interested in her current relationship, but a couple of days later she realized that she didn't want it any more the way it was. All of a sudden she felt she could let go of the relationship without feeling ashamed or hurt. She began feeling that she was worthy of love, and with this recognition she was able to let her partner go. She realized that he was not giving her love and he was not open for it.

Sometime later, she saw an interesting profile on a dating platform, and she immediately felt attracted to him. Normally, when looking at profiles, there was kind of a stereotype in her brain, and she had a strong pattern of dating men that would be approved by her environment. But this time she looked at this different profile, and for the first time in a long time she thought perhaps she should just start to trust her intuition. So, she followed her heart and got in contact with the man.

And this time everything started differently. For the first time she openly addressed her fears and how she tended to close up when there was too much intimacy. But at the same time, even without noticing, she opened up and let go a little bit. And he was sensitive enough to ask her, "Are you still ok that I am here, or do you want me to leave?" By connecting with her heart, she felt that she wanted him to stay. She addressed her inner struggle and didn't close up, and that was a big step for her. Before we started working on her energy, she would have closed up and let him go.

I really love how this one session changed her (love) life. She opened her heart to somebody who was fascinated by the same sports and interests, and there were so many things they could share together.

ENERGY EXERCISE #29 – Tear down your heart walls

Every one of us has already been hurt and upset, so most of us do have heart walls – they are a kind of automatic protection mechanism. But these heart walls can keep you stuck and make you relive certain situations over and over again. They separate you from your greatest inner strength, weaken you and, depending on the age and thickness of the heart wall, block you from feeling emotions and facing life, love, people, situations… and yourself. The strongest heart wall is the heart wall towards yourself. This wall does not allow self-love or self-forgiveness.

But your heart is brave. You don't need these heart walls anymore. Do you want to live your life open-heartedly again? Are you ready to dissolve your heart walls? Then *do* this powerful Energy Exercise!

During the Energy Exercise it is important that you acknowledge your heart walls. Just be aware that you built them up and that they are there. If you are interested, you can identify the situations and destructive beliefs behind them. And it might be helpful for your mind to let go of these old patterns, but most of the time you don't need to know these details to dissolve your heart walls. This Energy Exercise is all about healing the reasons and injuries behind your heart wall.

01 Stand up and close your eyes.

02 Bring your attention to your heart and connect with your heart.

03 Ask yourself if you are having a heart wall.

04 If you get a YES, ask yourself if you can dissolve your heart wall right away or if you need to let go of something first.

05 If you need to look into something first, check with the Wheel of Energy Blockades what you need to heal, clear out or let go of in Chapter 11.

06 Heal everything according to the Wheel of Energy Blockades that is keeping you from letting go from your heart.

07 Then say: "I now dissolve and let go of my heart wall," and visualize that the wall falls down into tiny pieces.

08 Check with the Standing Method if your heart wall is gone.

09 Ask yourself if you are having more heart walls.

10 If YES, continue with Step 4 accordingly.

Do you feel the difference? You are now much better connected with your heart, and yourself! Do you sense it? Well done! The goal is to build a connection to your heart so that you can better perceive your heart voice in everyday life again and let go of any resistance.

Dive into your flow again!

ENERGY EXERCISE #30 – Opening your heart changes everything

So now that you have been brave enough to remove your heart walls (isn't it a great feeling without them?

Scary but great, right?), the next step to liberate your heart is to open it.

A common mark of a closed-hearted person is their tendency to push away painful emotions. But you may remember, we already spoke about how this is not the best way to handle them.

Before you are ready to open your heart, let's see if there are any (emotional) wounds that need healing. How do you find out? Of course, with the Standing Method. Try the following Energy Exercise:

01   Stand up and close your eyes.

02   Inhale through your nose and exhale slowly through your mouth.

03   Connect with your heart by bringing your attention into your heart.

04   Ask yourself if there is a wound in your heart that needs healing.

05   If you get a YES: Imagine a powerful beam of white-golden light flowing right through you – from your head to your feet.

06   Connect with this powerful energy and send it into your heart to heal your wound.

07   Realize how your body tilts forwards.

08   Imagine how your heart heals and starts sparkling with this bright white-golden light.

09   Stay in this position until your heart is healed and your body tilts back into a straight position.

10. Feel whole, loved and safe, and stay there for a moment.

11. Check with the Standing Method if this wound is healed.

12. Ask yourself if your heart is 100% open.

13. If you get a NO, set the strong intention to open your heart to 100% and let everything go that prevents this. You can also find the cause behind it by going through the Wheel of Energy Blockades from Chapter 11.

14. Check with the Standing Method if your heart is 100% open.

In the course of your life, you have probably experienced at least a couple of heart wounds, but they all need healing. So, if you feel up to it, continue with the Energy Exercise right away. You can heal all wounds that are ready to be healed right now. Some may show up later because you maybe don't feel ready to look at them at the moment. Or something has to be healed or let go beforehand. But that is ok. It is a process that is liberating but sometimes takes its time. Just give yourself the freedom to look at issues when they show up. I have been working on my heart opening for a while, but even now, every once in a while, something pops up.

The good thing now is that you have this Energy Exercise to work on issues and let them go, which is empowering to know. In your daily life you can practice open-heartedness by giving yourself permission to cry. Crying is cathartic and is so good for you. Even, you might remember, in the supermarket…

Our heart yearns for novelty, for unfolding, and for adventure, but as you have learned, your brain lives by the rule: safety and comfort first! This means that there is a part of you that is not interested in going all out for your needs and dreams, because it might simply be too dangerous and too exhausting for it.

There will always be a voice in your head saying: "You don't need to do it. You managed well so far. What do you want to get out of it?" However, healing all these wounds and letting them go makes such a difference in your life! It feels amazing! Not only for you, but for everyone around you – and they will notice it, too.

ENERGY EXERCISE #31 – Your heart-brain coherence

If you want to boost your energy, productivity, and wellbeing, you need to try this exercise. It is all about the coherence between your heart and your brain.

So, what does it mean when your heart and brain are coherent? It represents this special state when your thoughts, emotions, and intentions are unified. When your heart and your brain are coherent, you will experience tremendous unity and peace with yourself and a deeper connection to those around you. Research shows that when you shift into a coherent state, the heart and brain operate synergistically, like two systems that mesh into one. This harmony is called coherence. The HeartMath Institute in the US has done a lot of research regarding this topic.

If your heart and brain are in coherence, it is reflected in the pattern of your heart's rhythm. This shift in your heart rhythm increases the coherence in all your physical processes. It results in a physical state in which the nervous, cardiovascular, hormonal, and immune systems are all working efficiently and harmoniously together. It is the highest level of function, where all your body is working together on an optimal performance level. Doesn't this sound like perfect health and heaven on earth to you?

You're right. And you will immediately feel a difference when doing this powerful Energy Exercise:

01 Stand up and close your eyes.

02 Put your left hand on your heart and your right hand on your forehead to connect your heart and your brain physically with each other.

03 Take a deep breath in and ask your body to create coherence between your heart and your brain.

04 Notice that your body tilts forwards.

05 Keep this position until your body gets back upright again.

06 Check with the Standing Method if your heart and brain are coherent.

07 Notice the change in your energy and your well-being.

My book mentor Cassandra tried this Energy Exercise, and she absolutely loved it. She gave me the feedback that she felt more in tune, quickly and easily. And Hannah, one of my clients, had the following experience. At first, she didn't feel anything. Just two minutes later it felt like her head was going up in an elevator and became much lighter (similar to a freefall in a rollercoaster, but in the other direction and without the grindy feeling in the stomach) – then all of a sudden her frequency took another massive jump. It is an amazing feeling. How do you feel?

This internal coherence is the state of being highly ordered, organized, and efficient. In a coherent system, all the individual parts are operating in harmony and *virtually no energy is lost*. I just love this aspect! It is a state of least effort and maximum benefit, because all the individual components are working optimally together rather than against each other. When your heart and brain are in coherence, it transmits waves of healing and transformation throughout your entire system, including your brain.

The key to integrate more heart-brain coherence in your life is: *appreciation!* Yes, you read that correctly. It is all about appreciating what you are doing and achieving, and appreciation for others. It is as simple as that. But is it really simple?

Yes, it is. There are easy things you can do to show your appreciation and to include it in your life, like thanking the lady that serves you at the coffee shop. Or you can even tell someone, "I appreciate your support today. I appreciate the time that you have spent with me to explain this." It is all about telling others that you appreciate what they are doing, or even what they are doing for you. This is very powerful, because appreciation is nice to hear, isn't it?

How often do you appreciate yourself for what you are doing all day long? How often are you celebrating what you have achieved? Just be honest with yourself. Nobody is ever going to know – unless you share. But you don't need to share. You just need to change!

When you feel appreciation, you are tuning into the good qualities of someone or something. It is both the awareness and recognition of those pleasing people, situations, and things – as well as the gratitude and thankfulness you foster for those experiences. This is a special state which you know intuitively, if for no other reason than because it feels good. Appreciation literally causes your energy to shift and expand. It puts a hard stop on the incoherence of the stress response. When you shift into a state of appreciation, you take your attention off what you don't have and put it on all that you do have. And in doing so – research has shown – you produce an extremely coherent, entrained, balanced, and harmonious energy field within the heart that is then radiated throughout your body.

It doesn't only sound amazing, it IS amazing! Integrate the Heart-Brain-Coherence Energy Exercise into your daily life and see how your appreciation of what you do and of what others do changes!

Your Energy Insights

There is a center in the body where love and spirit are joined, and that center is your heart. It is your heart that aches or fills with love, that feels compassion and trust, and that seems empty or overflowing.

Your heart provides you with emotional and intuitive guidance to help you direct your life. So, begin and end each day with two minutes of appreciation of the miracle of being alive and all the joys life has to offer. Free your heart and let love flow into your life and your energy!

If you feel stress or notice that you don't feel well at all, bring your heart and brain back in coherence. In this state you feel so much more powerful, and you have all the energy you need right away. Being in the state of coherence enables you to see opportunities and solutions you haven't been aware of before and also gives you the energy to execute them. It's a bit like having rose-colored glasses on.

Chapter Thirteen

Don't give the energy vampires a chance

PROTECT YOUR ENERGY

"Energy is contagious, positive and negative alike. I will forever be mindful of what and who I am allowing into my space."
Alex Elle

Don't give the energy vampires a chance

To show you how important it is to set energetic boundaries, I'd like to share this story from my first job. Most mornings, I would be looking forward to getting into the office to continue working on my project. It was really exciting. However, after some time I started noticing that no matter how good I felt in the morning, as soon as I sat at my desk, I felt angry. Sometimes I felt really bad, but I didn't understand where this was coming from. It was weird, because I liked going to the office and I enjoyed my job and making an impact with this project – at least, most of the time.

At that time, I didn't know anything about Personal Energy Management, so it took me a while to understand what was going on. But it turned out to be as simple as this: To get into my office, I always had to go through the secretaries' office, where the Chief Executive's two assistants were sitting.

And after a while I realized that the problem was not me. This was not MY bad mood! It was the bad mood of the two assistants sitting in the other office. Without realizing it, every morning when I walked through their office, I took on THEIR bad mood and their energy. What a bummer! Have you ever experienced something similar?

This is just an example to show you that each of us has our own energy. In Chapter Six, you may remember I showed you that you are made up of a physical body and a much larger energy body.

Your energy body can be the size of your body or the size of a city, depending on your personality, situation, training, and mood. You have probably experienced how sometimes a person enters the room and everybody turns around, because that person has such an energy that everybody notices and can feel it right away. These people have charisma, and we all know people like that. But they are rare. They probably have an energy body like, maybe, a small city.

Just observe yourself when you are not feeling well. Your shoulders are hunched, you are making yourself smaller, and so is your energy body. And if you're feeling well, your body has an upright position, and you are sparkling – then your energy body is big, too.

If you imagine this in practice, this also means that your energy body and those of the people around you always intermingle with each other. There is no other way, but many people are not aware of this.

This intermingling cannot be prevented but it is actually something beautiful, because it emphasizes once more the phenomenon of quantum physics that we are all energy and energetically connected with each other. Just the thought of this helps me to understand that we never need to feel lonely, because energetically we can always connect with somebody. And if you are aware of your energy and how to use it, you can deal with many situations in a completely different way.

Do you know your boundaries?

Human beings were biologically dependent on belonging to a tribe in earlier times, which protected them from starving, freezing, or being eaten. Adaptation and restraint were necessary for survival. As a child, you were dependent on your parents, and very few of us learn to distance ourselves from them, despite this dependence.

Maybe you are a person who manages her or his physical boundaries quite well, especially through body language. But managing your energetic boundaries is not always so easy – precisely because you can't see them. And it gets even more complex with emotional and also spiritual boundaries; I'm sure you get what I mean.

Boundaries distinguish what you are and what you are not, but it can be a challenge to recognize what is really you and what influence someone else has on you. Because your energy body is always intermingling, it is sometimes difficult to delineate. That's why it's so important to be aware of your own boundaries and to know them.

If you have often experienced that your own boundaries were not respected, you probably not only find it difficult to separate yourself, but even have trouble perceiving your own boundaries. Without energetic boundaries you lose contact with yourself, so it becomes difficult to feel what *you* want and what *you* need.

Setting boundaries means:

01 **Creating the right distance from other people. A distance at which you feel comfortable. This distance is not fixed. It depends on which situation you are in and who is around you. It can change within minutes. And this can mean physical as well as energetic or emotional boundaries.**

02 **Telling or showing other people when they are getting too close and bringing them back to the appropriate distance after crossing a boundary.**

03 **Knowing what you want and being able to express that.**

Just as your physical body is protected by your skin, you can imagine that your energetic body also has an energetic "protective boundary". Cyndi Dale wrote an entire book about this, *Energetic Boundaries,* in which she explains: "Strong and flexible energetic boundaries allow us to communicate who we really are to the world." From her perspective, maintaining energetic boundaries is extremely important for both your health and your personal integrity.

Think of yourself as a house. Your home is your personal space: you wouldn't allow anyone to just walk in, would you? So, how would you feel if someone just walked in your front door, went to your refrigerator, and started eating? You'd say no way, right? But if you don't maintain your own personal energetic space, you allow the thoughts, beliefs, emotions, and energies of others to overtake yours. And you even adopt some of them as your own!

Are any bells ringing? If so, it's time to explore a whole new awareness of boundaries. Boundary setting is not just about speaking your truth or setting limits. It requires a more subtle, different kind of awareness. It involves separating your own energy, emotions, and thoughts from those of others. It means declaring a boundary around your own sense of self and maintaining that dividing line.

Now that you can see what good it does to be able to set boundaries, this very powerful Energy Exercise will show you how to set your own. I wish I had known it when I had to deal with the energies of the secretaries at my first job. At that time, I didn't know how to protect my energy, nor how to use it in a positive way.

ENERGY EXERCISE #32 – How to set your energy boundaries

The basis for this Energy Exercise is to declare your own personal energetic space. What is you, and what is from someone else? This is precisely why it is so important to be aware of your own energy body and its boundaries, and to radiate and communicate that.

Being able to set boundaries doesn't mean becoming selfish and saying no to everything. It means being able to say no when you want to say no. Saying yes because you mean yes creates more closeness and trust. If you can express your wishes and needs in relationships – and this applies to partnerships just as much as to work relationships – you can also accept the other person with his or her needs. In the end, this makes relationships more cordial and loving.

You no longer have to draw walls around yourself, you no longer have to express "stay away from me" through facial expressions and body posture, you no longer have to make yourself invisible so that others don't get too close to you. You can allow more closeness because you now have the ability to set boundaries and stand up for your own needs. You just set your energetic boundaries. And after you have set the boundaries of your energy body, you will feel much more centered and protected.

Do you know where your energy boundary is? Is it close to your body, or is it further away? How does it feel for you? Imagine your energy body is about an arm-length from your body on all sides – like two meters around you. This size is comfortable for most situations, giving you enough room to breathe while maintaining a strong presence energetically.

Get started right away and learn how to set your boundaries with this Energy Exercise.

01   **Stand up. Place your feet hip-width apart.**

02   **Rub your hands in front of your body.**

03   **Move your hands apart and feel the energy between your hands.**

04   **Move your separated hands upwards above your head.**

05   **Say or think: "I set my boundaries here!"**

06   **Move your hands down, right and left from your ears and say or think: "I set my boundaries here!"**

07   **Move your hands down, right and left from your shoulders and say: "I set my boundaries here!"**

08   **Move along your body and continue to set your boundaries until you reach your feet.**

09   **Check with the Standing Method if your boundaries are set.**

How do you feel after setting your boundaries? Do you feel more centered and protected?

Did it change something physically in your body? Some of my clients experience that the tension in their shoulders relaxes. Isn't that fascinating?

Notice how comfortable you feel when you claim your space energetically! If you are adventurous, you can even imagine the color of your energy body. Does it look like a rainbow, or is there a color that is dominant?

If you practice this often, you will find that you feel safer, more stable, and stronger in almost any situation. It prevents the tendency to take on the emotional content of others, clarifies your own thoughts and feelings, and allows you to respond more effectively to relationships.

When you do this Energy Exercise every morning, you will realize that your energy body is becoming more stable and balanced. You consciously set your energy boundaries. Sometimes they can be closer to your body; sometimes you will set them further away, giving you more space. You can also adjust this according to whether you are at home and spending time with your family or out in public with complete strangers. People around you will notice the difference and address you accordingly. Just try it out! You will be amazed!

I learned from one of my clients, Sabrina, that this Energy Exercise can work in other ways. She told me that she was often in the uncomfortable situation that people were running or bumping into her, on the streets, in shopping malls, or just at work when she used the elevator.

This changed completely once she started doing the setting boundaries Energy Exercise. She told me that people started passing around her, and this running into her had stopped. It seems that people had been running into her because they had felt her energy body when they were already pretty close. She was so grateful that this was now over.

See? Personal Energy Management is all about adapting the Energy Exercises to your personal needs.

Letting go of external energies

The more you work and manage your energy in everyday life, the more you realize that not everything you feel actually belongs to you. This is referred to as external energies. Because, as I've already mentioned, your energy body intermingles with those of people around you, you can pick up and absorb their feelings, thoughts, and even habits. Sometimes these energies have been in your energy body for a long time without you even being aware of it.

You can recognize these external energies that have been taken over, for example, by this:

01 **You have recurring feelings of guilt, fear, anger or sadness, without actually knowing why.**

02 **You are constantly in conflict with certain people and cannot pinpoint exactly what it is that triggers you (e.g. men, women, bosses, friends).**

03 **You struggle with constant exhaustion, fatigue, weariness and restlessness and can't find any concrete reasons for it.**

These external energies make you feel as if you are being controlled by others and block your own power. So, if you all of a sudden feel strange, like being really angry or tired, or really feel bad or whatever even though nothing has happened, you have most probably picked up energy from somebody else that doesn't belong to you. It is all about raising your awareness and that you just realize:

01 **Is this me?**

02 **Does this belong to me?**

03 **Is this coming from somebody else?**

And then you check with the Standing Method whether you are having external energies. You can check for example by asking: Is this really me?

ENERGY EXERCISE #33 – Release external energy

Obviously, it is extremely important that you free yourself from these external energies and release them. Energy is neutral *per se*, and these energies are not bad, but they just don't belong to you. That's why it is so important to let them go and send them back to where they belong.

Right, let's check if you have some external energy.

01   **Stand up and close your eyes.**

02   **Check with the Standing Method if you are having external energy.**

03   **When you get a YES, proceed as follows:**

04   **Set the strong intention to clean and send back all the energies that do not belong to you to the person or place where they belong. Hold this intention to release all these energies.**

05   **Imagine that all these energies in your system are leaving your body and going back to the person or the place where they belong.**

06   **Check with the Standing Method if all external energy has been released.**

Set the very strong intention that you are sending this energy back where it belongs. Imagine that it flows out from your body, from your energy into the Universe, to the person or place it belongs. The rest you let flow into the ground, back to the earth. Feel the difference that has been cleaned in your energy. Do you feel so much lighter now? Repeat this Energy Exercise every evening before you go to bed.

Sarah, one of my clients, is using this Energy Exercise every evening after work but before she leaves her office. She doesn't want to take any of the energies, thoughts, and emotions from work back home to her family. So, she started a little ritual to finish her working day. Before she gets out of her office, she sets the intention that every single thing that belongs there, stays there. She cleans her energy before she leaves, so that she has all her energy for her children when she is at home with her family. I found this an amazing idea and now use it for myself as well.

And this is exactly the point of Personal Energy Management and the Energy Exercises I am sharing with you. This is just a beginning to get you started. But please experiment with your energy and the Energy Exercises. Adapt them to your own needs, mix them, and develop your own Energy Exercises! They are only here to serve *your needs*. And as you and your energy are special, you need to adapt them accordingly. Find your own way of working with your energy!

Want to protect your energy?

Based on Energy Exercise #14, you know how to recharge your energy. Imagine now that others can feel your energy as well. This can make you very attractive to certain people who want to take part in your energy. That can be the downside of having too much energy.

So, the paradox is – on one side the goal is to expand your energy body, to make it stronger and bigger; on the other hand, this means that your energy reaches out further and, hence, mingles more with people around you. For example, when you are standing in your kitchen, imagine if you expand your energy body up to five meters (meaning 10 m in diameter), part of your energy body is already in your neighbor's living room. Or imagine yourself on a bus or any other public transport, or walking through a busy shopping center. You definitely don't want to pick up everything around you, right? That's why on top of setting boundaries, you need to have some more protection.

Therefore, it is important to protect and shield your energy, and the simplest way is to set up an energy protection bubble.

ENERGY EXERCISE #34 – Set up your Energy Protection Bubble

Setting up an energy protection bubble[8] is something really powerful. It is a little bit tricky to set it up, and that's why I will give you some details before you get started. When I set up my first Energy Protection Bubble, I got it wrong. Let me explain, and please read this thoroughly before you set up your own.

I don't know if you remember the Grounding Exercise #15 in Chapter Seven. In that Energy Exercise you imagined that energy roots grew from your feet into the Earth, and an energy light beam connected you from the top of your head with the Universe. Do you remember?

When you set up your Energy Protection Bubble, you need to make sure that you can keep this connection with Mother Earth and the Universe by making a hole at the top and at the bottom of your bubble so that your energy beam stays connected. Not considering these connections was exactly the thing I got wrong. When I set up my first Energy Protection

---

8 I learned this technique from Stefan Klitzsch.

Bubble, I forgot about these essential connections and didn't make these two holes. What happened then?

Without the two holes, I lost connection with the Earth and the Universe. All of a sudden, I felt completely alone and isolated. It was such an awkward feeling, like being in a soundproof room or a vacuum. I can tell you it was not a nice experience to feel so cut off from everything. Luckily, my mentor helped me out of this situation and told me what to do to reconnect to the Earth and the Universe. So now I know how to set it up properly!

Let me put another little energy disclaimer on this here: I am writing this book to make you see, feel, experience, and discover that managing your energy is easy and can also be fun. But please be aware that whatever you do with your energy, it has an impact! Even if you don't see the energy, you will definitely see the result, so don't forget that!

There is one more thing you need to know. When setting up the Energy Protection Bubble, please set the intention that the hole is exactly as big as your personal energy beam, in a way that your energy beam is sealing off the two holes in your bubble and therefore nothing can get in. Your energy beam is always there and keeps you connected with Mother Earth and the Universe.

Now imagine a Protection Bubble around you and your energy body. At the beginning it's a kind of soft bubble adjusting around your energy body. And then imagine that this bubble opens up on top and bottom so that your life energy beam fits right through it and seals it off – at the top and at the bottom. Imagine now that your life energy beam fits perfectly through these two holes, leaving no space in between.

01 **Stand up. Place your feet hip-width apart. Close your eyes.**

02 **Imagine that you are connected with the Universe and Mother Earth with your infinite life energy beam that goes right through your body.**

03 **Imagine a protection bubble around you and your energy body.**

04 **Imagine a hole at the top and bottom part of the bubble.**

05 **Imagine that your life energy beam fits perfectly through these two holes without any space in between.**

06 **Test with The Standing Method if your energy protection bubble is optimally set up.**

07 **Notice how protected and safe you feel. You can also set the intention that the bubble protects you from all energies that do not belong to you.**

How do you feel after implementing your Energy Protection Bubble? Perfectly safe and protected?

When your Energy Protection Bubble is optimal, you can set an intention that it protects you from any external energies, all entities that do not belong to you or that do you any good. Only positive emotions get through it.

And the good thing about this is you only need to set up your very special Energy Protection Bubble once. That doesn't mean that it will stay the same for the rest of your life, though. On the contrary, you can adjust it, for example, so that it floats with the size of your energy body. And in special situations, you can customize it. This may depend on whatever you want protection from, like external energies or negative emotions from other people. You can just adapt it accordingly.

However, when you customize it, please always check with the Standing Method that the way you have adjusted it is best for you. The reason I point this out is the experience of another client, Jess. After she split up with her partner, she adjusted her Energy Protection Bubble accordingly. However, it seems that she exaggerated a little bit and made the Energy Protection Bubble too strong. All of a sudden, she got a headache, and it took her a while to realize the cause of it. Once she understood that the cause might be the adjustment of her Energy Protection Bubble and that she had also blocked out positive energies, she re-adjusted the settings of her bubble so that they were optimal for her current situation. She felt so much better afterwards, and the Energy Protection Bubble was then simply protecting her from external energies and emotions from her ex-partner.

If you have a strong energy field, you are not so easily influenced and also not so easily upset. A strong, closed energy body gives you strength and helps you to stay energetically with yourself.

I have had the Energy Bubble for a long time now and can clearly tell the difference. I have almost no external energies around me anymore and I feel calmer, more balanced, and somehow more protected.

Your Energy Insights

Your energy is your most valuable currency. The bigger and stronger your energy body, the more comfortable you feel. That's why it is so important that you protect your energy!

When you set your energetic boundaries, you distinguish yourself from others. It helps you to have a clearer picture about what you are and what you are not. It even helps you to better FEEL who you are. That's why it's so important to be aware of your boundaries and to know them.

When you set your energetic boundaries, you keep your energy with you. This means you need to worry less about where your energy always goes. It stays with you, and you will feel powerful and more productive and centered.

Chapter Fourteen
Your energy is the perfume to everything you do

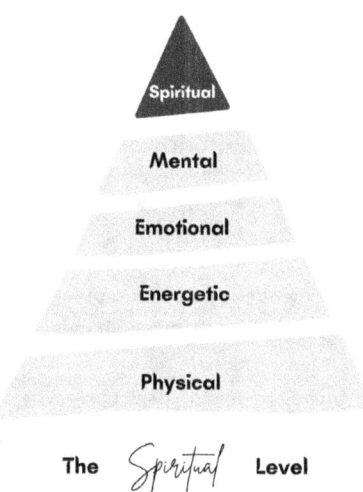

The *Spiritual* **Level**

As you can see, we are now reaching the top layer of the energy pyramid. This chapter is all about getting to know your spiritual energy. And yes, even if you are not a spiritual person, it's ok. Everybody has a spiritual energy, independent of whether you are aware of it or deny it; it doesn't matter. It just exists.

If you don't use this energy, that's fine. It just stays small and will diminish your all-over energy a little, because it is part of your energy, and it is part of you – whether you like it or not.

"Your energy introduces you before you even speak."
Jim Earle

HARNESS YOUR ENERGY

Get to know the top of the iceberg

The difference between spiritual and physical energies boils down to these definitions:

01 **Physical – It's the energy you use to do something.**

02 **Spiritual – It makes you up as a person. It's about your character as a human being.**

By learning more about your different energy levels and how they work, you will be able to harness their power to lead a happier and healthier life. Your spiritual energy can help you to reach your full potential in life. Every person has many unique spiritual powers. To know which powers you possess, be mindful of your inherent abilities. Observe the things that only YOU can do best.

For a very long time, I was not aware of my own spiritual powers. It was only when I fell into the deep dark hole that I realized that there is more behind being an electrical engineer. There is more behind just being a human being. I discovered my world of energy! Even though I still don't like that I only discovered my skills a couple of years ago, I am grateful that I *now know* that I can manage my energies in a more holistic way. I still cannot see energies, but I can feel them really well. There seems to be the right moment for everything. Whenever you are ready, your (new) superpowers will show up – if you haven't already taken them on.

I am glad that this journey opened up so many new ways of thinking, experiences, and broadened my consciousness, as I wouldn't want to have missed any of it. I have found so many skills I am capable of and that I was not aware of, like energy healing, and managing not only my own energies but also those of others. However, I am still on my journey. There is so much more out there to learn, explore, and apply. So, I'll keep going! How about you?

Understanding the importance of tapping into your spiritual energy can open up your awareness, new possibilities for self-improvement, and greater fulfillment in life. It supports you in having faith in life, having high-quality experiences, and to live a life with meaning. Also, being connected spiritually allows you to recognize imbalances in your life so that you can take actionable steps toward restoring them. Or it can help you to gain a new perspective. To cut a long story short, being aware of your spiritual energy can be a catalyst to living a meaningful life.

So, if you want true inner peace and harmony, then make sure you start exploring and tapping into your spiritual energy today.

Your energy is the perfume to everything you are doing

Did you sprinkle a little bit of your energy around you before you left the house today? No? Actually, you don't need to do that, because your energy is with you all the time anyway. And that is exactly the point.

Let's look at a situation in daily life, like going to the supermarket. I honestly don't like grocery shopping. For me, it is a pain in the neck – but yes, it needs to be done. Of course, I could order things online, but when it comes to food, I am very old-fashioned, and I want to see what I am buying. Actually, not only see them but also test whether the things I am putting into my basket are good for me. You might remember the zucchini story I shared with you in Chapter Five, when I wanted to test if this particular zucchini was good for me and my family.

If you haven't tried this powerful way of shopping until now – this is the moment to remember and just do it! By including not only the testing of the products but also your energy, you can give grocery shopping a completely new twist! You can start there and then expand it to everything you are doing. This is such a game-changer.

There are different ways you can accomplish any task. You can just get it done, or you can enjoy doing it. People will feel the difference, and probably the result will not be the same. Why is that? It is the energy that counts when you are doing something, no matter what it is.

Right, let's continue with doing the grocery in the supermarket. As I said before, there are different ways you can do it. With this example, I am outlining two alternative situations:

You are already late, you hate doing the groceries – especially now, as everybody else is doing the same. The supermarket is crowded. It is difficult to get through the aisles. You rush along the different aisles as quickly as possible and throw the items of your shopping list into your cart. You bump into other people who are in your way, as they are trying to do the same. Then, at the bread shelf, the person in front of you just takes the last loaf of your favorite bread. Now you are really pissed, because that happened last week, too, and you have been looking forward to eating your favorite bread for breakfast tomorrow. What a pity! Finally, you get to the cashier and need to queue up in one of the long lines. You get nervous because the cashier seems to be new and not as fast as you thought. You blame yourself that you haven't queued at the other line, which moves forward much faster. When it is finally your turn, you are annoyed, exhausted, and done with the day.

You enter the supermarket and greet the shop assistant with a friendly smile. You know her from the short chats you had with her when it was not such a busy day. You grab your basket and are looking forward to buying all these healthy vegetables and ingredients for your family. You enjoy doing the grocery shopping and move through the supermarket with ease. You look at the different products and then decide to take products that are in season. As the lady next to you puts her paper basket of apples in her cart, one apple falls out, and you pick it up for her. At the cashier, you get into the line and involve the woman in front of you in a little chat about a new product you have spotted in her basket as well. It is a nice conversation, and before you are even aware of it, it is your turn. After paying, you pick up your bags with your groceries and wish the cashier a wonderful day.

Of course, these two situations are a little bit clichéd.

Of course, I know that there is a huge gray area between these two extreme situations.

Of course, I know that you cannot always be in a good mood.

But I just wanted to paint these pictures so that you see the difference.

I just wanted to give you a flavor that sometimes life can be so much easier, yet we make it difficult for ourselves.

It is up to you to make the decision:

Every day you decide with which energy you want to go through the day.

Every day you decide how you want to feel.

Every day you decide which energy you want to put on and sprinkle around.

Most of the time we just forget that we can decide. But I am quite sure you remember a day when just everything works out the way you want it.

> You get to the bus stop as the bus is just about to leave, and the driver stops and opens the door just for you.
>
> Your boss praises you in front of all your colleagues for your valuable work.
>
> You turn on the radio and it plays your favorite song.
>
> You get compliments from strangers for your appearance.
>
> You finish writing a concept much faster than you had planned, and you know it is good.
>
> You get a letter from the tax authority, and after you have dared to open it, you read that you are getting a refund.
>
> You cook dinner for your family, and they are blown away because it tastes so delicious.

Every once in a while, we all have such "perfect" days, right? These are the special moments when you feel aligned with life.

But wouldn't it be wonderful to have more and more of such magic days in your life?

Everything that is in your life has something to do with you

The secret to such perfect days is your energy and the frequency you are vibrating on. Because everything that is in your life or is coming into your life has something to do with YOU. Yes, whether you like it or not – it has something to do with YOU.

When you contemplate your life, look at all different areas. Whether it is work, your relationships with your partner, family, friends, your health, your financial situation, what you have achieved so far – have a closer look, because all these six points have a vast impact on what your life looks like:

01 **What thoughts you think.**

02 **Which feelings you feel.**

03 **Which decisions you make.**

04 **What things you believe in.**

05 **What goals you pursue.**

06 **Which things you give power to.**

I know that's hard to understand and believe – but it is true. It is easier to blame others for things and people that are in your life, and to blame everything for not being able to enjoy the life you want. But that kind of thinking gets you nowhere.

Look at your life without any blame and take responsibility for why certain things and people are there. Because when you find the reason and cause behind it, you can change it. But ONLY then!

Empower yourself and take back the reins of your life. Find out why things you don't like are in your life then let them go.

The more you are managing your personal energy, the more you are shifting your energy, and the more you will feel, think, and even react differently. This will also lead to you having more energy. When you have more energy and your energy body is bigger, you feel more resilient, centered, and enjoy whatever you are doing much more. On top of that – because you are having more energy – you will also see and experience so many new opportunities you wouldn't have been aware of before. And you will even have the energy to take on many of these opportunities and leverage them. This will not only be noticed by you, but by everyone around you. They feel your energy in the same way as you feel the energy of others.

ENERGY EXERCISE #35 – What does this have to do with me?

Most of the time we spend so much energy trying to change other people, but this does not work. The only thing that can change is yourself and your energy. And if you don't change anything, you will create the same future over and over again.

So, if there is something in your life that you don't want to have anymore, make a deep dive into it. It can be anything that is bothering you, like eating too much sugar, spending so much time on social media, not doing any exercise, being in a toxic relationship, or having debts. Or all of them! Then continue with the following Energy Exercise:

01 **Stand up and close your eyes.**

02 **Connect with your energy.**

03 **Ask yourself: Does it have something to do with me?**

04 **If you get a YES, continue by asking: Is there a reason why it is in my life?**

05 **If you get a YES, ask: Can I change it?**

06 **If you get a YES, please check: Can I change it right now?**

07 **If you get another YES, then please continue with the Energy Exercise of the Wheel of Energy Blockades (see Chapter 11) and dissolve the reasons behind why this is in your life.**

Depending on what you don't want to have in your life anymore, you might need several rounds of this Energy Exercise to fully resolve it. Sometimes the issue can be complex, or you need to look at it from different angles. Sometimes it is also helpful to prioritize (please check with the Standing Method), then work through this topic according to the priorities.

And sometimes it takes longer, or we need to go right through it to be ready to let go of it.

Only when you work with your energy and find out what beliefs you have about yourself (for whatever reason) can you change your energy, and therefore your life. You can also apply this Energy Exercise for working on things you are lacking, by finding out the reasons behind it and working on them in the same way.

Take on the responsibility for your life

When I look back at my life so far, there are some acquaintances and situations I am not really proud of. Things that happened to me, things I did and said. Do you also feel like this when you look back at your life? I guess this is normal. Life wouldn't be life if it was perfect, right?

However, we tend to sweep situations and things we are not proud of under the carpet. The main reason is the fact that we are embarrassed by them or even ashamed. The best thing would be that they had never happened at all.

Nevertheless, these situations, emotions, and thoughts are still with you – no matter how deeply you bury them. So, what is the best way to handle such situations?

What I have figured out so far is something I mentioned right at the beginning. I pointed out then that you won't like this thought, but nevertheless, I am going to repeat this powerful sentence here and let it sink in.

Like it or not: *Everything that is in your life has something to do with you!*

Which means, everything that is in your life, whether you like it or not, is there for a reason. This may be that you need to learn something out of it, or with somebody. I know it sounds hard – and I struggled with this for a while myself. But now I can see the truth behind it.

So, the next question is: What can you do about it?

The most straightforward way is to connect with your energy and find out why this thing or person is in your life, then solve or heal the cause. Then things should change.

However, I found out that things develop much more smoothly when you take on the responsibility that this thing or person is in your life. You are not a victim – it is there for a reason. Don't blame anybody else for it. The best thing is to take on the responsibility for everything that is in your life. Sounds scary? Actually, it isn't.

Taking on responsibility enables you to do two things:

It brings you back into the driver's seat. You are not the victim anymore but can decide how you want to live your life.

Once you have taken on the responsibility, you can change and transform and let go more easily.

I never believed that taking on responsibility would have such a huge impact, but let me show how it affected my life. About three years ago I did something that was recommended to me by my mentor, and I followed his advice, even though my energy told me not to do it. I did trust my mentor more than myself. So, I went for it – against my energy telling me the contrary.

About a year later – because of all the developments that occurred since then – I realized that this was wrong, I had made a big mistake, and it had had quite a negative impact on my life. So, I started working on solving that problem and mitigating the issues that evolved. I put a lot of Personal Energy Management and healing work into the process, but I didn't manage to fully solve it.

It took me a while until I realized that I had never taken on the responsibility for my decision. There was no need to blame anybody for this development (especially not my mentor). It was *me* who had decided to go for it, so it was me that had to live with the consequences. It was difficult for me to acknowledge this to myself. To accept that I made this mistake. That I was wrong. That I had messed up.

But it was not until I took on the whole responsibility for this mess that I could fully transform and heal it. It was really like magic. As soon as I wholeheartedly took on the responsibility and stopped blaming others, and even more importantly myself, for being so stupid, I could suddenly transform the whole situation that had been bothering me for more than two years. I could let it all go, and my whole energy shifted. I cannot tell you how liberating that was!

ENERGY EXERCISE #36 – Take on responsibility for your life

I understood that only by taking on the responsibility could I actually let the situation go and, even more importantly, decide what I wanted to have and do instead. It gave me the opportunity to start all over again and make it better, or at least different from now on. Taking on your responsibility for everything and everybody that is in your life is such a game-changer.

01 Stand up and close your eyes.

02 Think of a situation in your life when you believe you made a wrong decision, you made a mistake, you regret or blame yourself for doing.

03 Connect with your energy.

04 Decide that you take on the full responsibility for this situation. Take on the responsibility for what you did, or what you said.

05 Check with the Standing Method if you have taken on the full responsibility for this situation.

06 If you get a NO, go back to connect with your energy and set the very strong intention that you take on your full responsibility.

07 If you get a YES, decide to let go everything that is connected with this situation.

08 Check with the Standing Method if the whole situation is now solved to 100%.

09 Feel how your energy transforms and shifts.

Taking on responsibility for everything and everybody that is in your life is so liberating. This includes taking on the responsibility for all your emotions, beliefs, and thoughts. Because only when you acknowledge that it actually has something to do with you, can you empower yourself to change things. The key is to accept that it is good that it (whatever it is) is in your life, because it gives you the possibility to see it, then acknowledge it, and finally transform it.

I read a lot of books about personal development, and I thought I could transform my life just by reading about it. However, you can only shift things when you connect with them, look at them, and expand your energy. It is not about reading but all about doing. Your mind can only support your transformation process, but nothing actually happens until you work on your energy and expand it.

What do you want to take responsibility for and to shift?

Take yourself on as you are

Do you accept yourself as you are? Or are there aspects that you do not like that worry or embarrass you?

Self-acceptance is a basic pillar for self-love and healthy self-esteem.

Self-acceptance is taking yourself on as you are.

Self-acceptance requires understanding that you are not your actions and qualities.

It requires accepting that your actions, mistakes, and weaknesses do not define you. Accepting yourself also means taking yourself as you are, regardless of your achievements or the approval of others. It means showing self-compassion, whether things are going well or badly.

Always be yourself – no matter what others think of you! Managing your energy helps you to discover your true you!

ENERGY EXERCISE #37 – I take myself on as I am

Even though self-improvement is a good thing, it's important that you identify and focus on your positive qualities rather than concentrating on the ones you feel negatively about. Making the best of your life starts with accepting your imperfections. The way you choose to think, speak, and feel about yourself *is a choice*! You may have spent your whole life talking about yourself in a negative way but that doesn't mean you have to continue on that path. You can now take yourself on as you are with this powerful Energy Exercise:

01 **Stand up and close your eyes.**

02 **Tell yourself: I take myself on as**
   - How I am
   - Who I am
   - What I am to a 100%.

03 **I allow it.**

04 **Feel how your body reacts to it.**

Taking yourself on as you are doesn't mean that you cannot develop yourself, enhance your strengths, or improve your emotional intelligence and soft and social skills. Self-acceptance means doing it from a healthy and self-compassionate place, instead of starting with that constant dissatisfaction of believing that you are not good enough. Do your best every day and be proud of yourself for your efforts.

Accepting yourself the way you are makes you feel lighter and freer. In changing the way you think about yourself, you'll also change the way you see the world.

Shape your environment

The entrepreneur and author Jim Rohn once famously stated, "You are the average of the five people you spend the most time with." You have probably heard this quote before. And it is so true.

The reason for this phenomenon is the so-called mirror neurons. Because of these mirror neurons in our brain, we adapt our beliefs and behavior to our environment – we mirror it. Most of the time we orient ourselves on the people around us, those we spend our daily life with.

At the end of last year, when I did my annual review blog, I thought about this sentence and all of a sudden it occurred to me that my five people have completely changed within the last two years. There were five people that I always thought needed to be around me, so I met these friends, and went to conferences and parties with them. But since Covid, this has changed. Zoom made it possible to connect easily with inspiring people all over the world, which is what I am doing now. I invite international masterminds and quantum science discussion groups literally into my living room, and even drink a virtual coffee with a friend in Boston.

So, the good news is that these people don't have to be physically in your life. You don't even need to be in Zoom calls with them. They can also be role models who are a part of your environment through books, podcasts, videos, or seminars. If you listen to inspiring podcasts every day, they will have an impact on you. These people can also be deceased or just fictional. The important thing is that they inspire you!

Bottom line is: *our relationships are energetic*! They can drain or boost our energy. And most of the time, we are not even aware of it.

When you are shifting and transforming your own energy, it ultimately has an impact on your environment, because energetically we are all connected via the quantum field. So, shifting your energy doesn't only have an impact on your life, but also on the life of the people around you.

Ready for your transformation?

Are you ready to accept changes? Do you really want to transform and change your life? Very often I experience that my clients have a huge resistance (consciously but also subconsciously) to opening up to this new world of energy and everything that is possible within. So, there might be some energetic blockers keeping you stuck and preventing you from moving on.

I have to admit I was not ready to go down this route for a long time. At the beginning, I didn't even notice. It took me quite some time to find out by myself that I had a huge blocker in my life that led to the situation where I felt completely stuck – not only with my business, but also in other areas of my life, like my health and my relationships. It was so weird. I kept working on my energy, but these changes didn't have a real impact in my life. It almost gave me the impression that managing my energy didn't work.

It took me a while to realize that many of the things I transformed and changed energetically were not visible in my outer world, which I didn't understand. My understanding of the energy world is that as soon as you heal, transform, or eliminate the root of a problem, it should be gone, along with everything that is associated and connected with it.

But of course, I KNOW that managing my energy does work. So, I searched deeper and deeper – like Sherlock Holmes – for a hint or a clue to know where I needed to search further.

Finally, a very close friend who I asked to look into my energy and help me out, gave me the essential advice to check my ancestor lines. And it turned out that I had inherited some strong limiting beliefs and stored emotions that I had not been aware of.

From a logical and consciousness point of view, it is so absurd that your ancestors would want to keep you from living the best version of yourself. I hadn't even considered that they would want me to reach everything I was dreaming of by in some way sabotaging and blocking all these changes. How could this happen?

However, there were things I was continuously working on yet seeing no visible change in the material world. Take my Instagram account, for example. There seemed to be an invisible threshold of 800 followers. I tried different ways of attracting new followers, like Reels, going live for 100 days, more stories, you name it. I observed that I indeed attracted more followers, but as soon as I hit the 800, it kind of stopped. And even worse, over time I started losing followers until I was down to 750. For a year or so, I was more or less oscillating between these thresholds.

And although I was working on it energetically, this didn't change. And then I realized that there was something bigger and powerful blocking me and my development that I was not aware of. We all have these blind spots when we need external support. There are blockers which you think subconsciously and that are too difficult for you to look at, so you aren't aware of them and need somebody else to point them out to you. With my friend's suggestion, I suddenly understood that some of my ancestors weren't ALLOWING me to succeed. Do you remember the story I told you in Chapter Eleven where rats passed their fear of the scent of cherry blossoms onto their descendants? It was a similar situation in my life. Because some of my ancestors hadn't made things happen in their lives, they didn't want me to thrive.

By connecting with my ancestors, I became aware and felt all the fear, stored emotions, beliefs, and reservations they had from their own sad experiences. All their blocking emotions like rage, hate, sadness, and shame had been passed on over generations. From their point of view, they wanted me to be stuck, because they had not been allowed to live the life *they* wanted. They didn't find my way of living appropriate, nor did they want me to do energy work. Their way of thinking was stuck in the Middle Ages, yet we are living in the 21st century. Crazy, isn't it?

And if you are an empowered woman in the 21st century with big dreams, there's the chance that your ancestors, with their medieval mindset, will find that offensive is high, therefore I would like to raise your awareness that your ancestors can be part of the reasons why you feel stuck or do not thrive with your business and career.

So, what did I finally do to solve this tricky situation? By connecting with their energy, I felt all the fear I had from these changes. Fear of what would happen in my life if all of a sudden I became successful, and more and more people would notice that my process of Personal Energy Management was actually working. My life would be turned upside down, and that apparently unconsciously scared the hell out of me, even though my brain kept telling me I wanted that success and visibility.

When I looked closer, I noticed that deep inside I was not ready to be seen, to share my magic, and to inspire more people with what I was doing.

Working with the Wheel of Emotions – which you already know from Chapter Eleven – I sorted out this situation with some of my ancestors. And during this process I finally managed to let go of their emotions and beliefs, and my fears, and more importantly gave myself the permission to continue on my path. I realized that it was sad what had happened to my ancestors, and I understood their experiences and beliefs, but I envisaged living the life I want… whatever that might bring along. I didn't know what it was going to bring, but if I didn't try, I would never know. So, I allowed myself to let everything I had already transformed energetically to also materialize in the outer world.

What happened afterwards?

Needless to say, all of a sudden I managed to jump off this invisible personal threshold of 800 followers, and as you are reading this book you can see that I overcame and let go of the situation with my ancestors and allowed myself to step up. It is a journey, though, and I am still traveling along!

I am sharing this story with you to raise your awareness that some things are not as straightforward as our brain makes us think. You get to the point you want to more quickly when you trust your energy – it always shows you the way. You just need to listen! Now, back to you. What are those fears and beliefs that are keeping you stuck and are blocking your way to kick off and continue your journey? Your energy is inexhaustible. It consists of an infinite fund of vitality, health, and life force that you can always access. Sometimes, however, the access is blocked, and there can be various reasons responsible for this.

Please, don't be hard on yourself. Everything you have learned and achieved so far has helped you to understand your life and how to transform it. If you hadn't gone through this process, you wouldn't be able to acknowledge and transform it.

Finding out that it was not about becoming a different person but letting go of everything that was not me, was the biggest transformation I have ever experienced. I am still on this journey to find and meet the true me, but working with my energy has made me see myself from a different angle – and that indeed I am a sparkling diamond! I was not aware of this wonderful feeling before, and it gives me so much confidence and power in myself!

Connect with your energy and you will know, too.

ENERGY EXERCISE #38 – Allow your self to change

If you feel stuck in your life, or things are not moving along as you want them to be, there is something in you keeping you from thriving. I didn't dare to check it out for a long time – and that was time that I wasted! These fears and beliefs were not even mine, but passed on by my ancestors. And they were so far-fetched that it took me ages to finally track them down.

Be brave enough and have a look to see if there is something blocking you as well.

01 Stand up and close your eyes.

02 Connect with your energy.

03 Ask yourself if there is something blocking you or your development or holding you back.

04 If you get a YES continue: Ask yourself separately, if these blockers are beliefs, emotions, energetic arrangements.

05 Ask yourself if these blockers are yours.

06 If you get a NO, let it go.

07 Ask yourself if you need help to heal these blockers.

08 If you get a NO, continue to heal the reasons and roots behind it according to the Wheel of Energy Blockades.

09 Check with the Standing Method if you healed the blockers to 100%.

10 Ask yourself which new things you want in your life and activate them.

11 Ask yourself if there is anything else to do.

12 If you get a YES, find out what there is still to do.

As we are all unique, it is impossible to describe a detailed process of how to remove your blockers. But the idea behind this Energy Exercise is merely to give you an idea of how to approach it, work on it, and heal it. Maybe you will be unable to solve it the first time, but if you keep going, you should be able to transform it eventually. It is a process! And just by looking into it, you are already starting the way of change. Trust your energy – you will get the right information on how to proceed at the right time!

I know it sounds vague, but this is what I am doing as well. Sometimes we feel stuck on our own, then just sharing openly with somebody can completely change the energy. As they say: "all our growth happens in relationships". And that is what happened to me when I was writing Chapter Twelve. I had quite a blocker when I was writing it, then I had a session with my writing mentor, and we explored what was keeping me from writing. Within a few minutes, it changed the energy, and I said to her, "I can't wait to go and write it." Normally, our call would last around two hours, but this had only been about 17 minutes, and she suggested we end the session right then.

That energy exchange just fascinates me. I think it is not only connecting with the energy of the other person; it is more that all the information is in the quantum field. And then when you are connected to this other person, they can probably download other information that you can't. If you are maybe blocked, and they don't have this block, they can get this information for you and then kind of transfer it over. That information allows you to bypass your block because they have given you the information or solution.

Your Energy Insights

Sometimes you feel stuck with your energy work, and this usually means you have landed on a blind spot. You can check that easily with the Standing Method. Then check with the Standing Method about who would be the right person to help you. Sometimes it is just one little piece we are missing which prevents us from moving ahead.

So please, don't shy away from asking somebody to support you. It will save you so much time and energy, and it is such a smart move. Once you know what the problem is, you can then either solve it with the other person, or somebody else, or you can work on it all by yourself again. Just check with the Standing Method what is the best way to get out of this situation!

Chapter Fifteen

Your energy shows you your way

YOUR ENERGY IS YOUR GUIDE

"The secret of change is to focus all of your energy not on fighting the old, but on building the new."
Dan Millman

The power of managing your energy

This is now the time to wrap up everything you have learned and experienced so far by reading this book. It might have turned your world (as you have known it) upside down, opened up new opportunities and perspectives, and maybe even changed the way you perceive the world. The assumption that everything consists of energies and frequencies changes a lot by itself. At least, that's what happened to me when I started discovering the world of energy. It's amazing, and I still just love it.

So, think again about the *PEEMS* (Physical – Energetic – Emotional – Mental – Spiritual) Model. In a nutshell it represents that you don't have only ONE energy, but there are different layers and dimensions to it. To deal with these different layers, you have to know how to work with the five main components:

01 **How to get in contact with your own energy and being aware of it.**

02 **How to measure and recharge your energy.**

03 **How to deal with limiting beliefs, stored emotions and the importance of forgiveness.**

04 **How to let energy blockades go and clean your energy.**

05 **How to protect your energy and being aware that every action has an effect, for better or for worse.**

Personal Energy Management is the key to a new era in your life.

You can be low on energy, sad, desperate, or angry, but now you have enough Energy Exercises at hand to help get you out of this low vibration within minutes. And that opens up a whole new perspective for you.

You can now shift your beliefs and resolve energy blockades without meditation or talking with somebody about it. You can do this just by connecting with your energy, feel what is blocking it from flowing freely, then just let it go.

Your energy always shows you your way – you just need to follow!

Will this book save you from falling into a dark hole again? No, because that is just life! I want to be candid with you. Even with Personal Energy Management, there will be another dark hole at some point in your life and there is nothing we can do about it.

However, the beauty of energy management is that you will be able to realize you are in a dark hole, but don't have to stay there. You now know it's a temporary state and that you have learned lots of Energy Exercises. When you start using the Energy Exercises and integrate them in your life, the holes you fall into may not be that deep anymore. And it is easier for you to accept that there are holes you are MEANT to fall into as a lesson to learn and to develop yourself. This helps you to work your way out of them much faster. Why? Because you now know how to manage your energy.

Life is in constant change, and you will continue to have ups and downs. Because everything is energy, everything also comes and goes in waves. However, these moments are also important, because otherwise we wouldn't be able to enjoy and appreciate the ups. Every moment, no matter how difficult, always holds the possibility for growth. And with this book you are better prepared for the challenges of life.

What I love most about Personal Energy Management is that you are not a victim anymore. You are back in the driver's seat again when you take on responsibility for your own life! And this knowledge alone gives you the strength and energy you need to manage and keep going to live YOUR life.

ENERGY EXERCISE #39 – Your energy shows you your way

Lately one of my clients, Mira, said to me, "You know, I am a very rational person. But with you I have discovered that there is so much more to it. Thank you so much for sharing your world of energy with me." Needless to say, she made my day.

Please remember that your energy is with you at all times. You have all the information on how to move on in your life and where to go. So continue to connect with your energy, because it shows you what gives you energy and what takes energy from you. It shows you if what you are doing is force or power. It shows you if the people around you sustain or drain your energy. Visualize yourself sparkling with energy. What would that feel like?

01   **You see yourself sparkling full of energy.**

02   **You feel lighter and more yourself.**

03   **You see your future, milestone by milestone.**

04   **You trust in all the miracles that will come your way.**

05   **You can make better decisions.**

06   **You can communicate better.**

07   **You establish habits in your daily life that give you energy.**

08   **You focus on the important things in your life.**

These are the benefits you can experience once you become friends with your energy. And now that you have reached this point in the book, you know how to do it.

Trust yourself and your energy.

Trust your energy that guides you safely on your path.

By doing that, you will suddenly see ways, possibilities, and solutions that were previously hidden from you. Your energy lets you see what's important to you.

Think of your energy as having a spotlight that shines a light only on certain information, and leaves everything else in the dark. Your energy shows you your way. Matter is always organized by energy.

Claim your energy to live your life

Do you sometimes close your eyes and imagine that you are living the life you were planning to lead? Do you remember when this last happened to you? A split second when you were relaxed and felt the bright side of life? Maybe during your last vacation?

These dreams can be simple, like happiness, good health, or financial independence, finding a place to call home, someone to love, or a path to inner peace. Perhaps it's just a life with less pain, heartache, or loneliness.

Unfortunately, most of the time life gets in between us and our dreams. In the busyness of modern society, with all the different obligations for your job and business, your family, maybe looking after aging parents, it is almost impossible to focus on what you really want. The goal is more about keeping all the different balls juggling in the air. This costs a lot of energy and can sometimes be overwhelming.

So many things can pop up and keep you from going after what you really want:

You spend hours at your job.

You catch a cold.

You start a relationship.

You take care of your family.

You do the groceries.

You clean your place.

You end a relationship.

You move to another place.

The list is endless and exhausting. It's no wonder you don't have enough energy for a better life. Most of the time we don't even have enough energy to maintain the life we have.

The truth is, because of our busy lives, we don't have enough energy to chase our dreams. But here is another secret I want to share with you: Achieving your dreams is an energy game.

I am quite sure you have probably experienced this already with a small wish. For instance, you are going out with your spouse or a friend for dinner. You are hungry and want to get there as soon as possible and wish that you will be able to find a parking spot right in front of the restaurant. And what happens? You drive down the street, and exactly in that moment somebody is pulling out from a parking lot – right in front of the restaurant you are going to. Just wonderful. You are very happy and feel lucky.

I don't have a car anymore, but when I had, I always asked my children to wish for a parking space at the place we were going to. It was a fun activity, and the girls always loved it. They loved it because almost every time it worked out the way we wished for. Why is that?

01 **Small wishes – little energy:**
For small wishes you need only a little energy, and most of us do have sufficient energy for small wishes. And usually it is for something that isn't important, so there is a fun and playful factor associated with it. It is entertaining, and no harm is done in case it doesn't work out. But interestingly enough, this is the main reason that it does work most of the time. It is the ease and joy associated with it.

02 **Big wishes – lots of energy:**
This is exactly the issue. In order to make a bigger dream happen, you need more energy. Actually, lots of energy. Most of the time, we don't have that amount of energy, and that's why fulfilling bigger dreams takes longer. We need to build up the appropriate amount of energy. On top of that, you need to focus on it for a longer time, which can be tricky as well. And last but not least, big wishes are something you really want. That's why we often lose the playfulness and joy by working on them, which makes them even harder to achieve.

03 **For making a wish come true, you also need different kinds of energy.**
a. Energy for setting the intention.
b. Energy to follow up by taking action. To take steps to make it happen. To become reality.

That's why it is sometimes easier to make excuses for failed dreams.

As you are reading this book, you can see that I finished my dream. For me it is still a kind of magic that this is actually happening. When I started writing the first chapter, I held the vision of this book being in your hands, making a difference to your life! And this vision was what helped me to keep going when writing proved to be challenging, when I felt stuck and overwhelmed because so many things in my life and in my corporate job got in the way. This vision (and of course my Personal Energy Management) helped me to overcome these obstacles.

When I first started, I did not realize the amount of energy that it takes to write a book. I really had to focus my energy, my time, and my attention on it, which wasn't always easy. My deputy resigned at work, so I had to take on most of her tasks on top of my own (which were already a lot). I also wanted to support my daughters during their busy and demanding exams. So, the only time when it was peaceful, and I could follow the flow of writing, was late at night. I'd sit outside on our little terrasse with all the beautiful plants, writing and imagining you reading this book. And how it might change your life created all of this beautiful and powerful energy – maybe you can feel it right now. I put a lot of energy into the book when writing it, but I also got a lot out of it.

If I can do that, you can do the same! And you can do it even better. Your life takes place here, in this moment. Don't look for your energy, your creativity, and your positive feelings in the past, and don't postpone them until tomorrow. Feel and live them NOW!

01. **Live your life full of energy – it is waiting to be lived by you.**

02. **Appreciate your powerful energy.**

03. **Don't let yourself be stopped by anybody. Especially not by yourself!**

04. **Give yourself permission to live your life.**

05. **You deserve all this new energy in your life.**

06. **You just need to allow it to happen.**

That's why it is so important to claim back your energy and your power. Actually, you need to claim back everything that is part of you and belongs to you, like:

01. **Your skills.**

02. **Your magic.**

03. **Your space.**

04. **Who and what you truly are.**

Please always remember that you have everything inside you to create exactly the life you dream of. Learn from your setbacks, work on yourself, and go your way with the certainty that life is always for you. Go ahead and dream, but make your first dream the gift of energy. Your future self will thank you for it. Make the decision here and now to take your destiny into your own hands.

Money is not the most important currency of our time. Your personal energy is. All the noblest dreams in the world mean nothing if you don't have the energy to pursue them.

ENERGY EXERCISE #40 – Visualizing your new You full of energy

You can do this short visualization exercise daily, right before going to sleep, or right after you wake up. In these phases of the day your brain is relaxed and in a state of alpha brainwaves.

01 Close your eyes. Inhale deeply in and out. Connect with your body. Become aware of your breath and allow yourself to enter a state of deep relaxation.

02 Now remind yourself how you feel when you are full of energy. Imagine that you feel the energy with every cell of your body and with your mind.

03 Imagine that your energy radiates from your heart like white-golden light rays of sunshine.

04 Imagine that your energy radiates through your whole body in each cell. You can feel that you are full of white-golden energy. You feel so light and empowered as if you are almost flying.

05 Imagine how your life would be if you would spend every day full of energy.

06 Ask yourself: What would be different? What would a typical day in my life look like? How do I feel? What does my life look like when I am completely in my energy, when I am in harmony with myself?

07 Imagine becoming one with your energy.

08 Breathe deeply in and out and return to the here and now.

Repeat this visualization exercise regularly and begin to live this new energized version of yourself in everyday life. Think like this new energized you, act like this new energized you, and thereby change your identity.

Have fun with it!

Why energy management is so valuable

This story just happened recently, and I found it so valuable that I want to share it with you. We spent a wonderful long weekend on Reichenau Island in Germany, which was really beautiful and relaxing. Only one thing didn't work, and that was the internet. There is only one transmitting station on the island, and that seemed to be a problem. However, we didn't allow it to spoil our vacation and got all the information we needed in different ways, like talking to people and checking out the timetable at the bus station. Being spoiled by having information available on your mobile phone all the time makes you a little bit lazy, yet there are also other ways to find out what you need.

On our way back, we met an older couple at the train station. He was complaining loudly that Germany is apparently one of the richest and most advanced countries of the world, but they haven't managed to provide internet to all areas of the country. I was standing next to him, and I told him I was really impressed that he was so familiar with his mobile phone and the internet. He explained that he was now 82 years old and had traveled the world but became really angry when traveling around his own country was made so difficult for him.

We spoke about whether he really wanted to let something so trivial as not having the internet spoil his wonderful trip, because it doesn't make sense to get angry about something that you cannot change. This was the first time I felt so grateful for all the energy work I have been doing for a while now. I realized that I am much more resilient against sudden changes or obstacles; it helps me to stay focused, and things that I cannot change hardly trigger me anymore. And if you trust your energy, you always find a solution anyway.

I know that the internet issue was only a triviality, but most of the time these little things can spoil our days. It is not the big things; it is often the small things that throw us off balance. But Personal Energy Management makes you feel stronger and at ease with yourself. Even if something brings you off balance, now you know that you have all the Energy Exercises to bring you back into stability and harmony. You might not notice immediately, but sooner or later you will. You won't understand why other people get upset, annoyed, or triggered by certain things, because you are vibrating on much higher frequencies and are above these energies. What a relief!

This is something I wish for you: That with any emotion, belief, or situation you are letting go, you are coming closer to *your real you*. This process will give you more energy, more self-confidence, self-love, and resilience. And that is priceless.

Just make it happen!

Your Energy Insights

Do you know why I am so passionate about Personal Energy Management? Because it is liberating me. And it is liberating you. You will still be able to remember your past, but the emotions will not be attached to it anymore. You will no longer be triggered by events of the past. That is what I find so wonderful.

If you don't have enough energy, you are missing out on many opportunities. Because you just don't see them. Research has shown that low-energy people have only limited perception. Their perception is limited because they are running on power-saving mode.

With unlimited energy you experience evolving possibilities for success, relationships, health, and well-being. And best of all, you have the energy to execute them. Your whole life changes when you have more energy.

Chapter Sixteen

This is the end? No, it is the beginning

UNLEASH YOUR ENERGY

"The only way we can change our lives is to change our energy to change the electromagnetic field, we are constantly broadcasting. In other words, to change our state of being, we have to change how we think and how we feel."
Dr. Joe Dispenza

This is the end? No, it is the beginning

I am often asked the question: "How do you manage to always be so optimistic, happy, and full of energy?" I always take that as a beautiful compliment, and every time it makes my heart sing.

You understand this now, because you know the story behind it.

You know that it was not always like this.

You know that I still have my fears and my stored emotions.

And that I am still working on managing my energy.

You know that I am still stuck in certain areas.

You know that I am still not where I want to be.

You know that I am not leading a perfect life.

You know that Personal Energy Management is an ongoing process.

However, you also know that with everything I let go, that with everything I am working on energetically, I have one less frequency that doesn't influence me and my life anymore. It brings me one step closer to my unique frequency and my real ME.

And you know what? Once I started, I just could not stop.

I am fascinated by all the things I discover and learn, and I just cannot imagine a life without managing my energy.

Despite all the progress and changes I have made to my life, I still clearly remember the special moment in 2017 that changed my life, as if it happened yesterday. I didn't want to be in this dark, black hole anymore. Enough was enough. I couldn't stand it any longer. I needed to find at least one way to get out of there. I felt very strongly that something needed to change.

I was standing at the seaside, looking up at the stars, when I decided: *Today my life is going to change. This change is going to start right now. It just has to.*

I didn't want to continue with the way I felt, so I connected with one of the stars and decided to leave everything behind. I decided that I wanted to start all over again, with a different life.

And that was when I discovered the world of quantum science and I decided to start working on and with my energy. I didn't know where the journey would take me, but I knew I had to try something different.

This was the beginning of a completely new life, full of energy. But there were a lot of obstacles in between. Because:

Nobody wants to face her or his fears.

Nobody wants to look into traumas.

Nobody wants to be confronted by stored emotions.

But:

You do want to liberate yourself.

You do want to live YOUR life.

Right? It is all about just getting started.

To decide to change something in your life.

To decide to change what and how you do certain things.

It is all about asking yourself this powerful question: *Do you love yourself enough to start working on your energy?*

That is the key that can change a lot in your life!

You can't change your life unless you change your energy!

Be aware of what you have achieved

I don't know what kind of person you are. Did you do the Energy Exercises right away, while reading the book? Have you already experienced that your energy has shifted? That something has changed?

Or are you someone who wants to read a book straight to the end and then pick some of the Energy Exercises to try?

Or you might be someone who just thinks, "What an interesting lecture," then puts the book on a shelf?

All of this is fair enough.

But now you have reached the final chapter of this book. So, acknowledge yourself for the great work you have done while reading this book! It's time to celebrate!

Whatever type of reader you are:

I want to congratulate you in any case.

I want to congratulate you that you have read the book this far.

I want to congratulate you for everything you have achieved so far.

This may be that you experimented with some of the Energy Exercises.

This might be that you thought about some of the Energy Exercises.

This might be that you will remember the book in a couple of months or years and *then* start working with the Energy Exercises.

But whichever is the case, you do now know a lot about how to manage your energy!

Throughout our time together, amongst other things you have come to understand how to:

01 **Communicate with your body.**

02 **Make conscious choices.**

03 **Be aware of your energy.**

04 **Clean your energy.**

05 **Recharge your energy.**

06 **Protect your energy.**

07 **Let go of limiting beliefs and stored emotions.**

08 **Forgive yourself and others.**

09 **Re-connect to your heart.**

10 **Manage your energy.**

Personal Energy Management is an important key to living a different and fulfilled life, so you can be really proud of yourself that you started this journey. And I am quite sure that by using the Energy Exercises you will have already achieved some shifts in your life. Thank yourself for all the changes you have made –

small and big ones. And appreciate how they are benefiting you.

Being a high energy person

I am quite sure you are not the same person after reading this book. I am very positive that your energy will have shifted, that you and your energy will have grown and certain things in your life have changed.

How do you feel right now?

What does your "new" energy and your life look like?

The feedback I receive for the biggest change and transformation is that you may feel lighter and more at ease with yourself. You feel lighter and more carefree because some of your ballast no longer holds you back and weighs you down. You are not threatened by your past anymore because now you know that and how you can deal with it. You are aware that it is actually a good sign that emotions and beliefs from you are showing up because this gives you the opportunity to let them go. This is the huge difference: You can only let go of things you are aware of and you have taken responsibility for.

Of course you let go of some of your energetic luggage, but the feeling of lightness stems mainly from perceiving life as easier and simpler. And this is true in the sense that with an overflow of energy you experience evolving possibilities for success, relationships, health, and wellbeing. You feel guided in the right direction and this may make you perceive life is not as hard on you as it used to be to me. When you are guided by your energy and you are following, everything is more in flow. And best of all: you have the energy to execute many of these evolving opportunities. Yes, you can choose! Your whole life changes when you have more energy.

With all your five levels of energy of the *PEEMS-Model* – namely the physical, energetic, emotional, mental and spiritual energy – operating optimally and in harmony, almost *no energy is lost*. In this state of least effort and maximum benefit, all energy levels are working optimally together.

You own your energy and that makes you feel more centered and connected with yourself.

This is also due to the fact that you now don't let energy vampires suck your energy out of you. That you set your energetic boundaries. You don't waste your time any longer on avoiding things and procrastination. *No, on the contrary: You take action!* Because you now know what to do and have the energy to do it.

This results in feeling powerful and in charge. It brings you back to being in the driving seat of your life, enables you to make powerful decisions and to take on again the responsibility for your life. You are more resilient against the challenges of life because, yes, now you know that you can handle them. There is this trust in yourself and your truths, and that makes the biggest change possible.

Are you experiencing that things you are doing have so much more power? This is because you feel the power of what you are doing which makes you more focused, efficient, and productive. You realized many things you are doing now create an impact. On your life and the life of others. With an overflow of energy you can support others more powerfully and in a sustainable way.

You ARE a high energy person now!

As a high energy person you don't settle for the life that's been handed to you. You don't just "fall into" the next job or relationship. You don't stay stuck in dead end careers or poor choices you made in the past.

As a high energy person you find passion and create goals. These passions and inspiration fuel your energy. Think about the difference between waking up feeling jazzed about the day ahead versus opening your eyes to a day you're dreading. You don't leave this to chance but make the effort to identify what excites you. You then create clear goals around your passions. You let these goals and passions draw you forward.

As Norman Vincent Peale pointed out, "The more you lose yourself in something bigger than yourself, the more energy you will have."

You now have all the energy to change your life!

Reflect on your most valuable insights

Working on your energy and expanding and increasing it, as well as raising your frequency, is one thing. Reaching a higher energetic level is another. But remaining on it (holding the frequency) is something different altogether. Hence, holding a high frequency – in other words, staying on a high energetic level – isn't always that easy. So please keep cleaning and recharging and managing your energy. You have already attained so much, but it is of the utmost importance that you keep going!

Take a few minutes to reflect on how your life has changed since you started reading this book. What do you no longer want anymore? What do you still want to be different? And most of all, what have you achieved and what can you celebrate?

Living your own energy is not a one-time thing that ends when you finish reading this book. Personal Energy Management is a beautiful, lifelong journey, so your true personal growth is just beginning now!

As you look back on this journey and consider all the Energy Exercises that you have learned, I am sure there are some that stand out to you. Maybe you want to reflect on the following questions:

01 **What Energy Exercise helped you to tune into your body and mind?**

02 **Where and how did you notice shifts and transformation?**

03 **Where and how do you find your energy, peace, joy and vitality in life?**

Make a note of your most important learnings from the book. This is important, as it is always useful to remember what you have learned. The most elementary step for integration is that you know as clearly and firmly as possible what you want to change and achieve – especially when you need a little inspiration or motivation in your daily life.

And here comes another disclaimer: I'm breaking my own rules here, but there's a good reason for it and I'll tell you why. I am going to ask you to write down five Energy Exercises you really enjoy and feel comfortable doing:

Write down 5 Energy Exercises you want to include in your daily routine:

01

02

03

04

05

Make a note of these favorite Energy Exercises, but not in a notebook. It's better to write them on a Post-It note, where you can see them every day. Perhaps stick it on the mirror in your bathroom (the classic location, I know, but it's very useful – despite what your other family members think about it), or you could place it on the door of your fridge (another classic) where you can probably see it even more often, or stick it on the lamp or something else next to your computer (I assume the most powerful way) to remind you that you need to do your Energy Exercises.

Why is it so difficult to establish routines?

"How do I continue and keep up that momentum?" That is one of the most frequent questions I am asked when finishing working with my clients.

And it's something I can really understand myself. During my life, I have attended so many great seminars and online courses for personal development and have been completely on fire every time. But then I would go back to my everyday life and immediately continue doing a lot of the same things as before.

One elementary factor for falling back into old patterns is our habits. And these habits run subconsciously. Change processes usually have these three phases:

01 **Enthusiasm:** you realize all the positive things in the change and start to implement it.

02 **Fear/Uncertainty:** usually after a short period of time the change feels uncomfortable and exhausting, but it is just unfamiliar and needs time, discipline and love.

03 **Integration:** the change becomes your new normality and a fixed habit in your life.

Evolutionarily, your brain's most important job is to keep you alive. As long as your life is not in danger, your brain wants to keep everything exactly as it is. If you try something new, your brain will always try to return you to old, familiar patterns of thinking and behavior. So, always remind yourself in such situations that the changes you are trying to bring into your life are not wrong, they are just unfamiliar.

In order to better understand and accept your thoughts and feelings, you should give yourself enough space during the change process. Be loving with yourself and, above all, connect again and again with your energy and let go of things that do not serve you anymore. Fall so much in love with your new energy that you can't help but keep going!

How did I integrate energy work and Energy Exercises into my life? The secret for me was to connect it with something I was already doing every day. Like in the morning: right after I get up, I clean and recharge my energy with Energy Exercises #11, #06, and #14. Afterwards, I ground myself (Energy Exercise #15), and I love the Chakra-Tapping Exercise #13. And then I set my energetic boundaries for the day with Energy Exercise #32.

At the beginning, the Energy Exercises were always the same, because I wanted to practice them, but once I knew how they worked, I used them every day. Now I follow my energy more and more and experiment with different exercises. Energy management should be fun; there should always be a playful component part of it.

When I started, I always did the Energy Exercises right after brushing my teeth. I linked them to an already existing routine to make sure I remembered to do them every day. I don't know about you, but I brush my teeth three times a day, and each time I also do the

Energy Exercises. I especially love doing the Heart-Brain-Coherence Energy Exercise (#31) after lunch, to focus and center myself to start the afternoon full of energy. If I have time, I also connect in with my heart (Energy Exercise #28) and my Inner Child (Energy Exercise #25) – just to check if everything is alright or if they need anything. And you could also integrate some Energy Exercises before you leave the office.

And it's the same in the evening. As I stand right next to my bed (before I climb into it – that is my reminder), I do the Energy Exercise #24 and Energy Exercises #11 and #33.

These tips should inspire you to develop your personal routine, even if you've never had one.

Needless to say, I use the Standing Method (Energy Exercise #8) hundreds of times during the day. When I get up, I sometimes use it to dress myself when I am not sure what to wear. I use it to plan my day and set priorities, who to call when I need information, when going shopping, planning dinner for the evening… The list is endless. In other words, the Standing Method is already part of me. My body now reacts sub-consciously when I am introduced to a new person and hence gives me additional information about their energy. It is just incredible.

Of course, I also use all the other Energy Exercises I shared with you: when something comes up that triggers me (Energy Exercise #22); some emotions suddenly show up (I can only recommend the 90 Seconds Rule from Energy Exercise #19, including Energy Exercise #20); and look behind where they are coming from (Energy Exercise #26); or I get stuck because a limiting belief is blocking me (remember Energy Exercise #21). Also make it a habit to connect with your heart (Energy Exercise #28) and allow your energy to show you the way (Energy Exercise #38).

Create your own powerful routines

After reading the book you now have lots of practical Energy Exercises that can immediately make you feel differently, think differently, and act differently.

Figuring out how to create a daily routine that works for you – and how to stick to it – can take some time. So, think carefully about what your needs are and what your perfect daily routine should contain. What are the routines in your life to which you could connect with one or several of the Energy Exercises?

Observe what you have to get done and when. To create a reasonable energy management routine you can stick to, you also need to be honest about your lifestyle, including your obligations and time-management abilities. Integrate your favorite Energy Exercises into your daily routines – like when you are waiting at the coffee machine in the morning or brushing your teeth. Then commit to the routine for at least 30 days.

Keep working on your energy! It is so valuable for yourself and your surroundings.

Now you can create your own personal daily framework with which you can grow and feel comfortable. I know what I said at the beginning – no writing.

BUT this is just to support you and keep you accountable. And that is my job, right?

So, when it feels right for you, answer these questions in writing.

01 How will you organize your daily routine from now on?

02 Which Energy Exercises are you integrating into your morning routine?

03 How will you design your morning routine from now on?

04 How will you design your evening routine from now on?

05 Which Energy Exercise are you integrating into your evening routine?

06 Which Energy Exercise are you going to use that give you energy during the day?

07 Which Energy Exercise are you going to use to support you in special situations?

Finally, one last and important point: Personal Energy Management is *not* dogmatic. What I have shared with you is a starting point for your own energy journey. The Energy Exercises I showed you are to give you a jump start, to get you going.

Make these Energy Exercises yours!

Play with them!

Experiment with them!

Tweak and adapt them!

Develop *your own* process with them!

This book is just the basis to get you started – the foundation of your Personal Energy Management. But Personal Energy Management is all about *your* energy. And that's what I love so much about managing my energy. There is no right or wrong. There is only you and *your energy*. There is only YOUR WAY!

Don't wait another five years. Deep inside you know your worth, but for too long you have been fulfilling the expectations of your family, your boss, your friends, the community around you. Now is the time to fulfill your *own* expectations. Once you have learned this, you can benefit from it for life.

You now know my personal energy story. Now it is time to write *your* energy story.

And if you still have any doubts that you can do this as well, please always remember that you are an energy being. Working with energy is natural to you. I know you can make it. I FEEL it. The only thing you need to do is to get started and do the Energy Exercises!

Do I use the Energy Exercises every day? Of course, I do!

I replenish my energy.

I clean my energy.

I protect my boundaries.

I look into the cause of situations when they trigger me.

I let go of my emotions when they show up.

Am I where I want to be? Of course not!

It is still a work in progress.

BUT I see that things are changing.

That my life is changing.

And that is what keeps me going.

And all the energy I have through my energy work kept me going during the whole process of writing this book. That's why I could finish what I started. That's why you are holding this book in your hands. The experience has been overwhelming at times, but I used my Energy Exercises regularly – and I finished, still smiling, with lots of energy!

Please always remember: No matter what you do or think, your energy is omnipresent. It is always in the center. That's why Personal Energy Management is the shortcut to personal development and being in the flow.

What are you waiting for?

Let me share one last story with you. I have always hated waiting. For me it was lost time, valuable time that I could have used completely differently instead of waiting for someone or something. I always wanted to get the most out of my time and tried to optimize it and squeeze in as much as possible. Queuing for the cashier at the supermarket was definitely a nightmare for me, and when a bus came late it stressed me a lot. I'm sure you see my point.

Of course, there was a story behind why I was so resistant when it came to waiting, and yes, I looked into it with my Energy Exercises and solved it. But that is not the point I want to make here. I didn't just solve it; I turned it around. I totally changed my perspective of waiting!

Now waiting for me is an unexpected, gifted time for – guess what – yes, for energy work! As you have already experienced, you can do most of the powerful Energy Exercises wherever you are, whenever you want and need them. That is the beauty of Personal Energy Management. You have everything with you – you just need to get started and do it.

And that is exactly what I do now when I need to wait. I check in with my energy, I recharge my energy, connect with my heart, and ask if there is anything I should let go. And then I just work on it. I am always amazed at how simply and easily it works. *Quasi en passant!*[9] Without anybody noticing it. And when it is your turn at the cashier, you are relaxed and calm and full of energy, so you can smile and wish her a wonderful day. She will be very grateful for that.

---

9   If you have read this far in the book and you're not aware that English is not my first language, then I have done incredibly well. But I would like to share with you a little phrase that we use, and I'd like to explain that to you. It means 'as a side effect'.

See, you don't have any excuses for not having time to do your Energy Exercises. You have your energy-gym always with you – that is the Standing Method. And anything else... well, just use your Personal Energy Management! It is your life, and you are responsible for it.

So go for it and rock it!

Your Energy Insights

CONGRATULATIONS!!! You have successfully completed reading this book. (Unless you cheated and skipped right to the end.) So it's time to celebrate!

Find something specific that makes you remember to do your Energy Exercises daily. Something that reminds you of your powerful energy. That is now YOUR energy talisman. It can be an object, such as a necklace or a stone, or even a movement, a pose, an affirmation, or a mantra. Feel what is right for you.

You could also use one of the powerful energy symbols at the beginning of each chapter. Pick one of the symbols (you can vary them every day) and draw it on your hand. This symbol now represents your unique energy that you have been working on while reading this book, and you can always connect to it when you need more energy, clarity, motivation, and joy. Furthermore, every time you look at your hands, you will get a reminder to do an Energy Exercise, to recharge your energy or to clean it, for example.

Every day is so valuable, because you are what you do every single day. Fulfilled and successful people are aware that every day counts and have a mindset, heartset, and energyset accordingly. They don't look for excuses why they could stay in bed half an hour longer today, or why they couldn't do sports after all. They focus their attention, and thus their energy, on the life they want to create and everything they want to do. They don't leave it to chance but make a conscious decision to do it.

Are you ready to do the same?

You now have everything you need to change your energy to change your life. So, connect with your power and let your energy shine. Or even better: Let you and your energy sparkle!

I am going to leave you with one more secret, but please don't tell anybody else: When managing your personal energy, YOU are the magic wand!

energy-on!

Your Magical Energy Guide

Cornelia

Chapter Seventeen

Your list of ENERGY EXERCISES

NR	ENERGY EXERCISE	CHAPTER
#01	Become aware what drains and sustains your energy	3
#02	Check out your energy balance	3
#03	Get your energy back	3
#04	What do you love and appreciate about your body?	4
#05	Make peace with your body	4
#06	Yes, Yes, Yes	4
#07	Discover your body's language	5
#08	Using your body as a pendulum? – The Standing Method	5
#09	Force or Power? The questions for your business	5
#10	Feel where your energy body is right now	6
#11	Clean your energy body	6

#12	Activate your Womb Chakra	6
#13	Chakra tapping and Recharging Method	6
#14	Recharge your personal energy	7
#15	How to ground yourself	7
#16	Measure your Bovis values	7
#17	Measure if food is suitable for you	7
#18	Find your optimal Bovis values	7
#19	How to apply the 90-Seconds Rule	8
#20	Let go of your emotions	8
#21	Eliminate limiting beliefs and activate supporting ones	9
#22	Let go of blame and accusations	9
#23	Forgiveness always takes place in the heart and not in the mind	10
#24	Ask for forgiveness, forgive others and yourself	10
#25	Talk to your Inner Child	11
#26	Clear your energy blocks and let go	11
#27	Find out your heart frequency	12
#28	Feel your heart and communicate with it	12
#29	Tear down your heart walls	12
#30	Opening your heart changes everything	12

#31	Your heart-brain coherence	12
#32	How to set your energy boundaries	13
#33	Release external energy	13
#34	Set up your energy Protection Bubble	13
#35	What does this have to do with me?	14
#36	Take on responsibility for your life	14
#37	I take myself on as I am	14
#38	Allow your self to change	14
#39	Your energy shows you your way	15
#40	Visualizing your new You full of energy	15

Acknowledgements

Personal Energy Management is so powerful. It makes you grateful, because you realize that it brings the right people into your life at the right time. And let's be honest, our relationships are the only thing that really matter in the end. Nothing else!

I guess, first of all, I need to thank my personal energy. I was not aware that I was planning to write a book. But all of a sudden, I had this strong impulse to do it. Can you imagine? Me?

And what a coincidence that right in that moment I spoke with the wonderful and inspiring Lizi Jackson-Barrett about how she wrote her bestseller. That was the first moment when I felt the urge to do the same. Before I was even aware of it, Lizi had already arranged a discovery call with her book-writing mentor, the magical Cassandra Welford. And I clearly remember the moment when I first spoke with Cassandra. She asked me, "What would you like to write a book about?" And without even thinking about it, I answered, "Energy is life!" And that was the very special moment when I knew that I was going to write a book.

Thank you, Lizi, for paving the way!

However, there is one person without whom this book definitely wouldn't exist – at least, not in the way that it is written now. And that is definitely Cassandra. Why? Because deep in my heart I am a scientist, and without her advice and guidance I would have written (or at least tried to) a scientific book. No kidding! But Cassandra showed me the way of heart-led writing, which made me create a completely different book – one that I enjoyed writing and hopefully you have enjoyed reading.

Thank you, Cassandra, for making me rediscover my love for writing again. Now I can't stop!

And a big thank you also belongs to the amazing Christine McPherson who edited the book. As you are aware, English is not my native language, so she had to deal with some quirky sentences and words and all the many prepositions I didn't use in the correct way. At least now I know some of them. Thank you so much for your patience!

Another important companion on my energy journey was my dear friend Johanna Köb. She was the first one to read the initial draft of the book. Without her smart questions and proposals for restructuring certain chapters, this book would definitely not be as valuable as it is right now. Once I incorporated her comments, it was the first time that I felt that this might become a good book.

Thank you, Johanna, for all your energetic and incredible insights.

I would like to thank all the wonderful women, especially from the Vitamin W group from Dame Tessy Antony de Nassau, who believed in me and supported my journey. I am incredibly grateful for all these amazing and powerful women around me, especially Renee LaPlante and Iwona Fluda, with all their crazy and heart-warming ideas. I just loved the exchanges with you.

A special thank you goes to the amazing Jessica Fabrizi and all her fabulous ideas and her amazing network. With her outstanding recommendations and her Personal PR expertise she outlined the way that this book becomes the visibility it needs. It is incredible how she thinks outside the box and makes things that are unthinkable happen.

When I was almost finished writing the first draft of the book, I felt the strong urge that I needed more support. So, I started my Facebook Group: Change your Energy – Change your Life! And I am very grateful for all the support I got there, if I was stuck and felt unmotivated, or I needed support on my bio, or which cover to choose – the list is endless! Thank you so much for being there for me and

accompanying me on my author journey. I strongly believe that the book wouldn't be the same without your frank opinions and recommendations!

And, of course, this wouldn't be the book it is without my two incredible daughters, Emma and Ronja. For the first two months I didn't even tell them that I was writing a book. I didn't tell ANYONE, because I felt insecure and didn't know if I would actually manage to complete it. I had so many doubts: Who am I to write a book? Do I really have something to say? But I knew that I had to work on all these limiting beliefs and let them go, because I wanted to be a role model for my girls. I wanted to show them that it is worthwhile to go for your dreams, no matter who tells you that it won't work out.

So, I wrote this book not only for YOU, but also for them. To show them that it is possible. To show them that it is worth the hassle. If you are having a WHY bigger than you, you can always find the HOW and energy to make it happen!

Thank you, Emma and Ronja, for always supporting me and my crazy ideas and believing in me. It means the world to me! I wouldn't be the same without you!

And finally, the biggest thank you goes to you – because you trusted me so much that you bought this book and started managing your personal energy. Thank you for your trust. I believe in you and your energy!

About Cornelia

Your Magical Energy Guide

Cornelia holds a PhD in electrical engineering, is an executive in the Swiss electricity regulator, and the founder of energy-on!

During her corporate career, she has held various management positions in several energy companies in Europe, all the while founding two start-up businesses. During her studies, she won scholarships at two American Universities and later worked at the Lawrence Berkeley Laboratory and in several energy companies in Europe. All these different touch points resulted in a vast global network.

As a Personal Energy Strategist under the brand energy-on! Cornelia empowers busy female exec-preneurs to use their full energy and thrive in their career and business. As an electrical engineer, working with energy and frequencies in all its forms comes naturally to her. She explains to her clients the world of quantum physics and shows them how to understand, measure, increase, and shift their energy. Managing their personal energy is the shortcut to reaching their goals.

When not working in the energy industry or with private clients, Cornelia can be found sharing her hacks in her energy-on! show, or interviewing inspiring energy workers and scientists. She also often appears on podcasts as a guest and at international conferences as a speaker.

Cornelia's mission is to make Personal Energy Management popular. She aims to inspire and empower YOU to feel the power of your energy and how to use it. Her passions include painting colorful pictures, swimming long distances, and visiting places with high energy. She comes from an Austrian family of entrepreneurs, which is why building companies and problem-solving is in her DNA.

Behind the scenes Cornelia loves talking to the many plants on her beautiful terrasse, singing ABBA songs, she eats any cake (not only vegan), and is crazy about drinking hot water because it has just so much more energy.

And psssst – another secret most people don't know about her. For three years Cornelia has run her own "Co-Ka" handbag brand, featuring her unique self-designed and hand-painted handbags at the Berlin Night of Fashion and the New York Haute Couture Fashion Week.

What a journey! And the journey continues…

A very special gift for you

Life can indeed be very stressful sometimes. Due to our busy lifestyles, we are often not even aware that we are stressed; we are already so used to it.

Leave stress behind and enjoy having even more energy with my very special complimentary ebook, *Energetic Stress Management – The NEW way of dealing with stress.*

You can download it here:

Contact Cornelia

I would love to hear from YOU!

As I already mentioned in the "Where do you begin?" chapter right at the beginning, I would love to hear your Personal Energy Management story and how changing your energy has changed your life.

And I am even more curious to see all the inspiring pictures where you are practicing your Energy Exercises. There is no more useless waiting, just quality time for you and your Energy Exercises. So, I can't wait to get all your cool ideas.

And if you'd like to find out more about me and what I do, check out my website:

https://www.corneliakawann.com/

My Facebook group to join my energy journey:

Change Your Energy - Change Your Life | Facebook

Don't miss out on my Podcast "The energy-on!" Show:

https://www.youtube.com/channel/Cornelia_energy_on

Follow me on Instagram, LinkedIn, and TikTok for more powerful Energy Hacks:

https://www.instagram.com/cornelia_energy_on/
https://www.linkedin.com/in/cornelia-energy-on/
https://www.tiktok.com/@corneliaenergyon

Come and take part in an empowering, uplifting, and inspiring exchange on how changing your energy can change your life.

I love to work with YOU and your energy!

Now that you have finished reading this book, you might be so excited about Personal Energy Management that you want to make a deep dive and join one of my programs. Or maybe I can support you in a one-to-one session – just you and me and our energies. You will find all the available programs outlined on my homepage.

And if you have any questions, please reach out to me at *info@corneliakawann.com*